How to Argue

Prentice Hall LIFE

If life is what you make it, then making it better starts here.

What we learn today can change our lives tomorrow. It can change our goals or change our minds; open up new opportunities or simply inspire us to make a difference. That's why we have created a new breed of books that do more to help you make more of *your* life.

Whether you want more confidence or less stress, a new skill or a different perspective, we've designed *Prentice Hall Life* books to help you make a change for the better. Together with our authors we share a commitment to bring you the brightest ideas and best ways to manage your life, work and wealth.

In these pages we hope you'll find the ideas you need for the life *you* want. Go on, help yourself.

It's what you make it

How to Argue

Powerfully, Persuasively, Positively

JONATHAN HERRING

Prentice Hall Life
is an imprint of

Harlow, England • London • New York • Boston • San Francisco • Toronto • Sydney • Singapore • Hong Kong
Tokyo • Seoul • Taipei • New Delhi • Cape Town • Madrid • Mexico City • Amsterdam • Munich • Paris • Milan

PEARSON EDUCATION LIMITED
Edinburgh Gate
Harlow CM20 2JE
Tel: +44 (0)1279 623623
Fax: +44 (0)1279 431059
Website: www.pearsoned.co.uk

First published in Great Britain in 2011

ISBN 978-0-273-73418-5

British Library Cataloguing-in-Publication Data
A catalogue record for this book is available from the British Library

Library of Congress Cataloging-in-Publication Data
Herring, Jonathan.
How to argue : powerfully, persuasively, positively / Johnathan Herring.
p. cm.
ISBN 978-0-273-73418-5 (pbk.)
1. Persuasion (Psychology) 2. Interpersonal communication. I. Title.
BF637.P4H465 2011
168--dc22 2010030249

10 9 8 7 6 5 4 3 2 1
14 13 12 11 10

Typeset in 9.5/13pt IowanOldBT by 30
Printed in Great Britain by Henry Ling Ltd, at the Dorset Press, Dorchester, Dorset

Contents

Part 2: Situations where arguments commonly arise

Introduction

Do you hate arguments and avoid them at all costs? Or do you just find that you keep losing them? Perhaps even when you win, somehow you feel it has all been counter-productive?

If so, this is the book for you. It will teach you how to argue well. You'll discover how you can get your points across in a clear and effective way. It will also help you to develop techniques so that you can respond to the arguments of others equally effectively.

Some people love arguments (lawyers and small children in particular). But most people flee them. Sometimes that's a good thing, but often it isn't. Avoiding an argument can mean that the problem simply goes on and is brushed under the carpet. The suppressed resentment can poison a relationship or fill a workplace with tension.

In this book we will look at more positive ways of understanding arguments. They needn't be about shouting or imposing your will on someone. A good argument shouldn't involve screaming, squabbling or fisticuffs, even though too often it does. Shouting matches are rarely beneficial to anyone. Instead, we should view the ability to argue well as an art and a skill.

The ability to argue calmly, rationally and well is a real asset at work and in life. It can sharpen your thinking, test your theories, get you what you want. In any case, it's impossible to avoid arguments. So you need to learn how to argue well. Arguments can be positive. A good argument between friends can be fun and enlivening. An argument can get matters out in the open so that issues can be dealt with and there are no hidden grudges. Sometimes an argument is necessary to ensure that we get what we are entitled to: if you never argue in favour of a pay rise, you might never get one!

Arguments should be about understanding other people better, sharing ideas and finding mutually beneficial ways ahead. Arguing has sometimes got a bad press. But that's because people often argue badly. That must stop!

“The aim of argument, or of discussion, should not be victory but progress.” Karl Popper

Arguing should lead to a better understanding of another person's point of view and a better understanding of your own. Many people go through their lives simply not understanding how anyone could be a socialist, believe in God, support fox-hunting, or enjoy French films. This happens because they've not discussed these issues with people with whom they disagree. They've not presented their views and had them tested by others. It's astounding how many preconceptions people have about those who are different from them. 'It's amazing, I met a Conservative Party supporter the other day and they were quite nice,' a friend once said to me. It's only by talking to other people who disagree with you that your own responses become clearer and you can better appreciate the views of others.

This book is in two parts. The first will set out what I call the *Ten Golden Rules of Argument*. These are rules that can be relevant in a whole range of situations: from arguments with a boss, to arguments with a partner, to arguments with your plumber. They'll even work if your partner is the plumber! In the second part I will look at particular situations where arguments commonly arise. We'll put the golden rules into practice.

Part 1

The ten golden rules of argument

In this part I will introduce you to the *Ten Golden Rules of Argument*. These will help you in any argument you come across. Once you have understood them you will be able to argue well with whoever you encounter. The golden rules apply to arguments anywhere: at home, at work, at play, or even in the bath!

Chapter 1

Golden Rule 1

Be prepared

Those Scouts are on to something. Being prepared for an argument is key to success. Sometimes arguments come out of the blue. But not always. It may be that you realize a difficult business meeting or conversation is going to take place, in which case being prepared is a real advantage.

What do you want?

Before starting an argument think carefully about what it is you are arguing about and what it is you want. This may sound obvious. But it's crucially important. What do you really want from this argument? Do you want the other person just to understand your point of view? Or are you seeking a tangible result? If it's a tangible result, you must ask yourself whether the result you have in mind is realistic and whether it's obtainable. If it's not realistic or obtainable, then a verbal battle might damage a valuable relationship.

Imagine you would like a pay rise. You have arranged a meeting to discuss this with your manager. Think carefully about whether this is a realistic goal. Is it clear the company is making cutbacks and all budgets are being drastically reduced? If so, the likelihood of getting a rise is probably nil and there's little point asking for it. But are there other things you can do to achieve higher pay? Is there a promotion you can apply for? Increased training you can do? Can you offer to do something extra for the company? Think through the options before you enter the room. Always enter an argument with a clear view about what you want at the end.

Framing an argument

When preparing your argument, spend time thinking about how to present your point in a logical way. Admittedly, logic has a bad reputation.

> "Logic is the art of going wrong with confidence." Joseph Krutch

People are often put off by references to logic. There is even suspicion that logic is some kind of clever trick to trip up those who are not 'trained' in logic. In fact, there's no magic to it. True, professional logicians have developed rules of magnificent complexity, but everyday logic is not difficult to grasp.

Logicians talk about a 'premise' and a 'conclusion'. A premise is a fact upon which it logically follows that there will be a particular conclusion. For example: 'I like all action films, therefore I like James Bond movies.' Here the premise is that I like action films and the logical conclusion is that I like James Bond movies. Sometimes several premises are needed to reach a conclusion. In a complex argument, a series of logical conclusions can be drawn from an initial premise. Consider this fine example of an argument:

> "[T]he evils of the world are due to moral defects quite as much as to lack of intelligence [premise]. But the human race has not hitherto discovered any method of eradicating moral defects . . . Intelligence, on the contrary, is easily improved by methods known to every competent educator. Therefore, until some method of teaching virtue has been discovered, progress will have to be sought by improvement of intelligence rather than of morals [conclusion]." Bertrand Russell

A good argument, then, is not just saying what you think but offering a set of reasons for it. Bad arguments will involve people simply repeating their conclusions to each other:

Getting it wrong

Bob: 'Men can't do washing up. They just aren't programmed that way.'

Marie: 'That's rubbish.'

Bob: 'No, men are just different from women.'

Marie: 'That's sexist – there's no difference.'

Bob: 'It's just obvious women have different brains.'

Marie: 'You don't know what you're talking about.'

This is typical of many arguments. All Bob and Marie are doing are repeating their conclusions to each other. There's no possibility of any progression. This is because they are stating their conclusions and not giving the reasons for their beliefs. If either were to say 'Now why do you say that?' or 'Do you have any evidence for that claim?' then progress could be made. They might be able to begin a useful argument through which each party could start to understand why the other person thinks as they do.

So if you're trying to make an argument that's convincing you need to start with some facts (premises) that the other party will accept as true and then move to a conclusion that must logically flow from the premise. There are two things you need to be confident about:

1. Make sure your facts (your premises) are correct.
2. Make sure that your conclusions necessarily follow from your facts.

Facts

We need to say a little bit more about facts.

Using facts

It should be obvious that facts are essential to many debates and arguments. Before starting any argument it's important you discover the information about it. You're going to lose an argument about the benefits of the European Monetary Union if you have only read a couple of blogs about it and are discussing the issue with a professor in economics. You'll lose an argument about a pay rise if you don't know what similar workers in your company and in other companies are earning. Arguing without facts is like trying to make a snowman with cold water.

Finding facts

Unless you are someone's parent, or are particularly well-respected, 'because I say so' isn't going to get you very far. You need to refer to facts to back up your argument. The internet is most people's first stop for information, although it's well known that this must be used with care.

Tip: Many search engines have a 'Scholar' button that will direct you to academic studies. These might be a more respectable source of information than the view of a blogger, but beware: you may come across some of the most impenetrable language possible!

It's dangerous to assume things are true just because they're well known. Here are some well-known assumptions that are simply wrong:

- Goldfish have a memory of only a few seconds. False: in experiments it has been found that goldfish can navigate complex mazes.
- Thomas Crapper invented the flush toilet. False: it was invented by Sir John Harrington in 1596.
- Shaving makes hair grow back quicker. False: it doesn't, nor does it make hair thicker or coarser.

Of course, libraries, newspapers, magazines and friends can also provide a source of information too. Make sure your source of information is respected.

Is the information reliable?

This is an important part of the task and needs to be handled with care.

- The source of the statistics can be key. The best source may be a group or organization that is respected by the person you're arguing with. If that's not possible, then an organization that is neutral or very well regarded. A study produced by a small pressure group on the dangers of eating too much meat is unlikely to be as persuasive as a report by the World Health Organization saying the same thing. So consider: Who produced the study? Was the group likely to be biased? Is it a respected body or a little-known pressure group?
- What source will most influence the person you're arguing with? If you tell a creationist what an atheist scientist has said they may be suspicious. However, give them a study from a scientist who is Christian and they may be more convinced. Otherwise it's easy for them to dismiss the study as 'biased'.
- With regard to citing statistics to support your argument, how large was the sample? When a study is undertaken this normally involves interviewing or testing a sample of people and generalizing from that. So if 100 people are interviewed about, say, whether they like Marmite and it is found that 38 do, we are told that 38 per cent of people like Marmite. Of course this does not mean that everyone in the world was asked, but the researcher assumes that if 38 per cent of the sample liked Marmite then it is likely to reflect the opinion of people generally. However, crucial to this assumption is the size of the sample. If you asked just two people if they like Marmite and one did, that would be weak evidence that 50 per cent of people liked Marmite. You couldn't assume that the views of two people would match the whole population! Generally the larger the sample the more reliable the survey is likely to be. If the study doesn't say how many people were involved, be suspicious. Be very suspicious.
- Another statistic issue: how representative was the sample? Always find out who was surveyed. If you interviewed only

those visiting a Marmite museum then it would not be surprising that a large number of people liked Marmite. Watch out particularly for groups who say 'of those who phoned our hotline, 86 per cent agreed that ...' If people contacted the pressure group for help they are likely to be sympathetic to the group's aims. You can't assume they are representative of all people. The best studies are those that sample a large cross-section of the population, and these results will better support your argument.

> A study found that 70 per cent of smokers surveyed had tried to stop smoking, and not one had succeeded. That sounds like terrible news for those trying to give up. However, the poll had only interviewed smokers. So it was hardly surprising there were no successes!

- Listen carefully to what is being claimed. Be especially wary of 'up to' claims. If the argument evidence shows that pollution levels have risen by up to 35 per cent that means that 35 per cent is the very top level the evidence indicates. The true average figure is not disclosed and may be much less than 35 per cent. Also beware of studies showing that people are 'possibly' and /or 'considering' something. A survey that showed that more than 50 per cent of people were possibly considering using air travel less hardly shows that people are flying less!
- Watch out for 'maybe' or 'don't know'. Consider a survey where people were asked, 'Should the UK leave the EU?' They were allowed to answer 'yes', 'no', or 'don't know'. Let's imagine 15 per cent of people say 'yes', 20 per cent say 'no' and 65 per cent say 'don't know'. You can present the last two statistics by saying 85 per cent of those questioned did not support the UK leaving the EU, or 80 per cent of those questioned supported the UK remaining in the EU.
- Be very careful of percentages. Take a (fictitious) claim that drinking coffee increases your risk of heart attack by 35 per cent. Such a claim may well send you heading for the nearest bar. But before you do, such a statistic is highly misleading. First, we need to know to whom the risk applies. Is the

> increased risk only for those of a certain age, or those prone to heart attacks, or for the 'average person'? Secondly, you need to think about what the risk of heart attack is in the first place. We could say that going for walks in the countryside increases your risk of being hit by an asteroid by 300 per cent, but you would probably not worry because the risk isn't high in the first place. So a horrifying-looking increase in a risk is irrelevant if the orginal risk is very low.

There are two lessons here. First, if you are going to rely on statistics make sure they are the best ones available: from a reliable source, with a large sample, with a clear conclusion. Secondly, if the person you are arguing with presents statistics ask some of the questions above. You might then explain why your study is far more convincing than theirs.

Explaining statistics

Don't assume that the more statistics you have the better. A few well-placed statistics can be more effective than a long stream of them, which will leave the listener drowsy and confused. Only the most hardened statistic-nerd can take in more than a couple in a conversation. If necessary you can always say: 'I have a lot of statistics I could use, but let me tell you these two.'

Present statistics well. It may be you're addressing people who are familiar with the use of them, but often people find statistics hard to grasp. It can be best to present them in as personal a way as possible. So instead of saying 'Twenty-five per cent of women will experience domestic violence at some point in their lives', it might be more effective to say: 'If you have a room of twenty women you could expect five to have experienced domestic violence.' Not only does that make the statistic easier to understand, but it has more dramatic impact.

Tip: If statistics are about money and you want to show how expensive something is, put them in terms of individuals. For example: 'If we took the money that it will cost to buy the furniture for the reception area and divide it between the people at this meeting, we could all afford a two-week holiday to Florida.'

It's an easy and all too common mistake to make generalizations: 'Everyone knows . . .', 'All illegal immigrants . . .' These overarching statements are simply asking to be refuted by an exception that shows that the statement is untrue. There are very few statements of this generalized kind that cannot be refuted, so avoid using generalizations.

All generalizations should be avoided – except this one!

Presenting an argument

A key part of preparing for any confrontation is not only marshalling facts and reasons but thinking of how to present them. Obviously this will depend a bit on whether the argument is part of a meeting, a conversation or a presentation. But the basic principles will be the same.

Make it clear what you're arguing for and why

It's always good to set out at the start what you're arguing for and why. Consider this opening of an argument:

'The company should support the proposal to purchase the building at 3 New Street. I will demonstrate three reasons why. First, doing so will generate a considerable profit. Secondly, we have a real need for more space. Thirdly, it will improve the public image of the company.'

At the very start the arguer makes it clear what they're arguing in favour of and informs the listener by giving evidence of the three facts that will establish the case. Similarly, at the end of the argument repeat what has been shown:

'So we have seen that adopting this proposal to buy 3 New Street will generate considerable profit. We are in desperate need of space and buying that building will sort that problem out. Thirdly, adopting this proposal will greatly improve the public image of the company. I urge you to support this proposal.'

Note that the start and conclusion have put the reasons supporting the argument in their simplest form. There is obviously much more that might need to be said in the middle, but start and end with the three key points you're using to support your argument.

Tip: There is a well-known rule: tell people what you are going to say; then tell them again; then tell them what you have said. This is often said. For a good reason: it's extremely good advice.

One benefit of repetition is that, simply, it drives a fact home. Repeating a point at least three time is a popular technique of advertisers. Once you have heard five times that a particular product kills all known germs, you start to believe it.

Summary

Prepare for arguments well. Make sure you have researched your facts. Choose carefully the key arguments you will rely on. Work out what are the basic points you want to make and how you will present the arguments.

In practice

Write down what you want to say in bullet points. Use the following structure:

- premise
- supporting facts/reasons
- conclusion.

Keep your notes brief, then speak them out loud, slowly, three times. Then when it comes to having your argument, whether with a doctor, your spouse or an electrician, you will be able to speak 'off the cuff' in a convincing way. Of course, refer to your notes if you find it helpful.

Chapter
2

Golden Rule 2

When to argue, when to walk away

I'm sure we've all had arguments where later we feel that it was just the wrong time and the wrong place. Knowing when to enter into an argument and when not to is a vital skill. Before embarking on an argument always ask yourself: is this the right place and the right time? Is it better to walk away and not have the argument at all, or to have it at another time and in another place?

Entering arguments

Think especially about the following:

- Could there be a productive outcome from this argument?
- Is it better to have the argument in private or with other people around?
- Do you have the information you need to make a good argument?
- Are you feeling emotionally ready for the argument?
- Is the other person emotionally ready to hear your arguments?

Let's consider these points separately.

Could this argument be productive?

There's little point in having an argument if no good to anyone could come from it. Imagine you were at a works party held to try to drum up new business. You introduce yourself to a distinguished looking man who soon informs you that he's the head of the local hunt. You are strongly opposed to hunting. You could enter into an argument over the morality of hunting, but it's highly unlikely that this would be productive. You're not likely to put forward any arguments he doesn't know already. In a party environment you can't give a long lecture on the evils of hunting. The argument is not going to get anywhere and you may even end up damaging the interests of your business. It's time to walk away or quickly change the conversation.

Or, imagine the family Christmas dinner and Uncle Geoff starts making some homophobic comments that you find objectionable. There may be a time and place to talk about the issues with Geoff, but Christmas dinner is probably not it. The end result of any argument is pretty predictable: you and Uncle Geoff will both get upset and the rest of the family will not be best pleased with you! Leave it for another time.

There are some people who are so emotionally committed to their point of view that they're unlikely to change it. You're unlikely to persuade someone in a single conversation that their religion is wrong. The most you might hope to do is create a doubt that they will want to explore another time.

Useful example

'What evidence would you need to change your mind?'

That's a telling question. If the person suggests that no evidence could prove them wrong, then you know you're dealing with a total fanatic. Walk away!

Never argue with a fanatic, it's a waste of time.

Private or public?

This can be an important issue, especially in a business context. You need to think carefully about it. Is this an argument best raised on a one-to-one level with the individual concerned, or is it better discussed in a group? There are several issues to think about:

- *Confidentiality*. If in the course of the argument you need to raise issues that are confidential (either about yourself or someone else) then you need to make sure the conversation is in private so you don't breach confidentiality.
- *Confidence*. Will you feel more confident if someone else is with you? Or if you're alone? If you want someone with you, who will it be?

- *Formality*. Would you feel more comfortable raising this in a formal setting, such as a meeting, or in an informal one?
- *Intimidation*. If you know that the other person can be aggressive and unpleasant, it's probably best to try to see them with someone else or in a public meeting. They're less likely to be bullying if there are other people around. If they are offensive then, hopefully, there will be people who can come to your defence.
- *Agreement*. Are there other people who agree with you? If so, your argument may be stronger if presented in a group with others supporting your viewpoint.

Do you have all the information you need?

At all costs avoid an argument if you're unprepared for it. As we saw in Golden Rule 1: having the key information at your fingertips is crucial. There should be no shame in saying: 'I need to think through the issues a bit more before giving you my opinion. Let's talk about this tomorrow.' It may be that in the course of the argument information was presented that you did not know about. Again it may be best to take a break. You may need to go away and read the study the other person was talking about or get some more figures.

Are you emotionally ready?

Arguing well requires time, attention and effort. Arguing when you're exhausted, emotional or hurried is normally counterproductive. Even if you're always exhausted, emotional and hurried, you need to try to choose a time when you are well prepared and in a good position to explain your arguments and listen to the other person's. A snatched conversation about a pay rise over the coffee machine is not likely to work. A discussion about where a relationship is going is unlikely to work well at 1 a.m.

Be particularly wary about entering an argument when you're angry. The temptation on hearing that a person has made a decision you disagree with is to rush off an angry email or run round to see them. Be very careful. Make sure you're right in your understanding. It can be very embarrassing to go storming into someone's office to complain about a decision only to discover that you've got completely the wrong end of the stick.

Is the other person ready?

The issues we have been discussing also apply to the other person. You may be perfectly poised for the argument, but the other person must be receptive to what you're going to say. Perhaps you need to give them some information to read first before talking to them. You might even want to give them a short document setting out your points and suggest meeting to discuss them. That will give the other person time to think through what you want to say and provide a considered response.

Think carefully about what time is best to discuss the issue. Friday at 4 p.m. might be a great time for you, but your boss might be exhausted and stressed out. Again, the key point is whether they will be able to pay attention to what you have to say and give time to listening to your arguments properly.

Useful examples

'This is a really important issue and we must discuss this properly. I don't think this is the right time to do it.'

'Shall we discuss this more tomorrow when we have more time?'

'Ah, that old chestnut. Well, we could discuss that until the cows come home, but maybe it would be more fun if you told us about your holiday in Skegness.'

Avoiding arguments

Do you find you keep having arguments when you don't want to? You can stop this.

Tip: You don't have to argue about everything you disagree with.

Is this really necessary?

First, before every argument, ask yourself: is this really necessary? Maybe you think you're surrounded by fools and incompetents. Even if that's so, you don't have to correct every fool you come across. Just let some things be. Remember some useful phrases to escape an argument:

Useful examples

'That's a really complex issue.'

'Wow, that's an interesting point of view.'

'Well, we could debate that until morning.'

If you really feel you can't let the person remain in their ignorance, it may be better to be non-confrontational. How about:

'I read a really interesting article on that the other day. I'll email it over to you.'

Is this an insoluble issue?

Many of the great issues that people love to debate are simply reflections of a broader disagreement between the parties. For example, a dispute over whether life begins at conception often in fact reflects a debate over whether or not people believe in God. Unless you have a lot of time on your hands (e.g. you're stuck in a broken-down train) you're not going to be able to argue all of the issues thoroughly. If you're not going to be able to resolve it, maybe it is better to leave it.

It may be that the issue is solvable, but the other party is simply immovable. They're committed to a particular view and whatever you say is not going to change their mind. In this case the argument is unlikely to be productive. Warning signs of this will be when they simply seem unwilling to enter a discussion:

'I just don't want to argue about this.'

'My mind is made up.'

Or even, as someone once said to me:

> 'Whatever you say I'm not going to change my mind.'

Be careful about assuming some principle of rationality. Many people's beliefs are just assumptions, not based on thought or logic. It's amazing how many people will adamantly support something without thinking about it. I remember a conversation many years ago with my grandmother, a fierce supporter of the Conservative Party. I went through a range of issues (e.g. education, defence) and on each she supported the policy of the Labour Party, rather than the Conservative Party. At the end I remember saying, 'But Granny, on every issue you support the policies of the Labour Party, so why do you support the Conservatives?' 'Because I always have,' she replied, and there was no easy response to that.

Know your buttons

Most people have 'buttons'. As soon as someone mentions a particular issue we launch into our tirade. Twenty minutes later our poor friend looks at us exhausted and says, 'Well, I guess I shouldn't have said that.' Once you know that you have an issue about which you feel so strongly that you are likely to overreact, *beware*. Realize your propensity to lose perspective, make sure you keep calm, and ask yourself the questions: is this the time, the place and the person?

Summary

Remember you don't have to argue about everything you disagree with. Often it is better to leave things be. If it's necessary to argue make sure you're prepared. Make sure it's the right time and place to conduct the conversation. If not, wait for another day.

In practice

Take a deep breath, ask yourself whether this is the right

- time,
- place, or
- person.

If it is, take another deep breath and begin. If not, walk away.

Chapter
3

Golden Rule 3

What you say and how you say it

In an ideal world all arguments would be decided on their merits and not their presentation. But we aren't in an ideal world. There's no getting away from the fact that presentation of an argument is crucial. Advertising is all based on persuading you to buy a product that you would not otherwise buy, and most advertising is the triumph of spin over substance. Many people have won arguments, based on bad grounds, because they've made their points well. And many people with good points have lost their argument because they failed to make their case attractively.

To regard an argument as simply an intellectual battle would be a serious error. Many arguments involve emotional matters as much as intellectual ones. Have you ever heard a great speech or a great lecture? It was probably not due to the intellectual power behind the arguments, but the emotional appeal made. Barack Obama won the American presidential election not really because of the intellectual appeal of his argument but the emotional appeal and his convincing delivery.

Presentation

So what can you do to make your argument as attractive as possible? Here are some pointers.

Clarity

It's a big mistake to think that the more complicated your argument the more convincing it is. Even the most difficult of issues can be boiled down to a few simple points. That's not a call for dumbing down. You may need to include some complex ideas, but nearly always you can return at the end to your few key points. If those you're arguing with don't understand the claim you're making or why you're making it you're unlikely to make progress.

It's well known that fraud trials can be difficult to prosecute. One reason is that they're easy for defence lawyers to defend. All you need to do is confuse the jury. Introduce a mass of complex financial information and a few jargon-filled experts and soon the jury feels lost. They certainly cannot be sure the defendants committed the crime.

The same is true in arguing. Baffle your opponent and you might persuade them that the issue is very complex. But you won't persuade them you're right.

Brevity

I've said this once already, but I say it again. Keep it brief. A useful guide is the 'postcard test'. Can you summarize what you want to say on a postcard? Unless you've been asked specifically to comment on an issue, you should limit yourself to three key points at the most.

Most people say too much when arguing.

It's better to make one point clearly than forty that leave the listener confused or bored, or most likely both. Remember, only one argument needs to work. So choose your best ones and make the most of them.

Focus on the question 'What do they *need* to know?' If you're telling people what they already know they will be bored. They will not want to put in the effort of listening to you if within your 15-minute rant there's not a single point they don't already know. I know it's tempting to make every point you can, but keep some in reserve. Listen to the response of the other person. Are they taking your three main points on board, or do you need to explain them? Are they nearly persuaded by your three points, in which case some of your secondary points may be useful? Are they very knowledgeable about the issue? In that case you need to tread carefully!

Enthusiasm

Be enthusiastic about your argument. There's nothing wrong in showing people that you care about the issue. In your arguments don't be aggressive, but do be positive and lively. If you come across as bored or disinterested, you shouldn't be surprised if people feel the same about what you're saying!

Get the start right

When you start your argument you want to get people seeing the issue from your side straight away. Lawyers know this well. Their opening speeches seek to influence the perspective from which juries look at the case:

Getting it right

Lawyer for the defence: 'This is a case of an innocent family man, wrongly identified by a confused witness, who has been brutally beaten by the police. You must stand up for the rights of the innocent.'

Lawyer for the prosecution: 'This man has savagely attacked a helpless grandmother in her own home. There is plenty of evidence against him. We must protect society against this menace.'

OK, perhaps they wouldn't put it quite like that, but you get the idea. A lawyer's opening is key because they want the jury to look at the evidence from a particular perspective. So if, for example, you want to argue that adopting a particular proposal will severely endanger the financial well-being of your company you want those you're talking to to look at all the evidence, asking 'What are the financial risks here and how will they affect me?' If you can get them to consider the proposal from that perspective, you will be well on your way to winning your case.

Burden of proof

This is a really important issue in arguing, but many people don't appreciate its significance. Consider a chair of a meeting who says this:

'Well, this proposal looks very interesting. Can anyone think of any reason why we should not proceed?'

By putting the question this way the chair has put the burden of proof on those who don't think that the proposal should go ahead. There's no need to make the case for the proposal, that's assumed. Imagine if the chair had said:

'Well, here's the proposal. Does anyone think a convincing case has been made to adopt this?'

Hence the approach to take in arguing in favour of buying a particular new car is to say:

'Give me one good reason why we should not buy this car.'

By saying this the assumption is made that buying the car is good. You might have said:

'Give me one good reason why we should buy this car.'

But that would put the burden of proof on finding good reasons to purchase the car. So in arguing, seek to steer the argument into asking why your point of view should not be accepted. That way sceptics will remain on your side, unless they're convinced that there is a good reason to go against you.

Threes

All good things come in threes. An exaggeration, maybe, but remember:

Snap, Crackle and Pop!

Advertisers often use trios. They know they work.

Numbering your points might sound rather formal but it helps the listener to see where you're going and helps them to remember. It also helps the listener realize there's a limit to how long they will need to listen to you.

'There are three main reasons why I think we should support this project. First ...'

This assures your listeners that you're not talking off the top of your head. You have thought through the issue and respect the fact that they have limited time.

Don't be one-sided

The temptation when arguing is to put just your side of the argument. Door-to-door salespeople will always do this. They'll list all of the benefits of purchasing the product and try to avoid you thinking about the disadvantages: most obviously the fact it will cost you a lot of money! And I'm sure we've all met pessimists who, with every proposal, consider only the down sides. A pessimist will greet every proposal for a holiday with:

'Well it may rain; the hotel may be awful; I'll hate the food and it will be so expensive.'

It's a wonder some people ever get out of bed! Or get in it, for that matter!

A really good argument, however, will seek to respond to those that may be raised by the other side. Indeed, if you're able to present the argument against your proposal and then dismiss it that can take the wind out of your opponent's sails. If they try to repeat the argument then it sounds as if they're being repetitious and those listening will already have a negative view about it.

There's obviously a slight danger here. If you go on too much about the other side's arguments you may start sowing seeds of doubt into your listener. You may even give opponents a good idea about arguments to use against you! I suggest two key rules:

- Don't raise a counter-argument unless you have a good response to it.
- Do raise a counter-argument if there's an obvious response.

Tip: Rebut your opponent's arguments in advance when you can. Otherwise don't mention them!

The use of humour

Humour can be very important in winning an argument. It can play an important role in getting people on your side. If you manage to start your argument with a good joke, people may be keener to listen to what you say in the hope that you might have another one! Laughter can unite your audience and help them associate themselves with you.

There are, however, dangers with humour. Two in particular come to mind. First, it might distract those listening to you. I'm sure we have all heard talks and at the end said, 'That was hilarious, but what was it he was saying?' That might not matter if you're just trying to get people to like you, for example, if you're standing for an election. But it does if you're trying to get a serious point across. So using jokes as light-hearted contrast is something to be encouraged, but don't overdo it.

Secondly, it is generally best to avoid 'cruel' humour.

Getting it wrong

'I don't know what makes you so stupid but it's working.'

'I'm a little busy right now. Could I ignore you some other time?'

Making an unpleasant remark about the person you're disagreeing with may generate a quick laugh, but it's unlikely to endear you or your argument to those listening and is certainly unlikely to mean you'll have a productive discussion with the person you're talking to. You want people laughing with you, not at your opponent.

Use emotional associations

Restaurants in America, it's said, charge on average 15 per cent more for dishes that refer to 'Mom's specials'. It's amazing how a homely analogy can make something ordinary appear special. That's true too of arguments.

For many of us there are words or images or smells that convey a host of emotions. Not for nothing do estate agents suggest brewing coffee or baking bread just before a potential purchaser comes round to look at a house. Advertisers pay huge sums for celebrities to promote their product and they think carefully of the associations that are drawn. With one figure you may immediately think of reliability and trustworthiness and so they're used to promote financial products. Another personality is associated with beauty or sexiness and so they're used to promote a perfume.

So when making arguments make use of positive associations. What association do you want with your argument? Are you wanting to appear ruthless? Kindly? Financially astute? Associate your argument with things that your listeners will associate with a particular attribute.

> **Useful example**
>
> 'This proposal is as short and sharp as Alan Sugar's way of sacking people.'

Think carefully about the words you use. As we all know, words can convey a loaded meaning. Tabloid writers know this well:

'PERVERT STALKS TWEENIES'

is a far more eye-catching headline than:

'MAN FOUND LOITERING OUTSIDE SCHOOL'

The use of words can be very important. When thinking about how to express your argument choose the words that express the case strongly. Sometimes an ear-catching phrase can win an argument more effectively than a hundred statistics.

The abusive analogy

This involves linking the argument of another person with something unpleasant. In other words one pours scorn on the argument of another, but by sprinkling it with wit the attack is more attractive. There's always a danger that in rejecting someone's argument forcefully you will come across as rude. This is unlikely to be productive for your relationship with the person you're arguing with or those listening. Using a humorous comparison can enable you to be rude about the other person without appearing mean! It must, however, be treated with care. Get it wrong and you can lose the sympathy of those you're talking to. For example, consider this description of a person's argument:

Useful example

'A speech like a Texas longhorn: a point here, a point there, but a whole lot of bull in between.'

Keep cool

It's crucial to keep cool. A certain way to lose an argument is to start shouting at the other person. I remember seeing a father once screaming as loudly as he could at his toddler: 'I'm a loving father, you must do what I say.' The aggression in his voice spoke more loudly than the content of his words.

Yet we all know that tempers are one of the first things lost in many arguments. It's easy to say one should keep cool, but how do you do it?

The first point to remember is that sometimes in arguments the other person is trying to get you to be angry. They may be saying things that are deliberately designed to annoy you. They know that if they get you to lose your cool you'll say something that sounds foolish; you'll simply get angry and then it will be impossible for you to win the argument. Notice how rarely politicians get angry. They know that appearing to lose their cool will cost them any appeal with voters. So don't fall for it. A remark may be made to incite your anger, but responding with a cool answer that focuses on the issue raised is likely to be most effective. Indeed any perceptive listener will admire the fact that you didn't 'rise to the bait'.

Tip: Be aware that the other person might be trying to annoy you. Be aware of the kinds of situations where you might get angry. Avoid them.

If you feel yourself getting angry, keep calm and focused on the issue. If the person has said something personal against you, ignore it.

Bob: 'You're a fascist racist. You're scum.'

Tom: 'Bob, look, we're talking about whether positive discrimination should be allowed. This is a complex issue. Now, I was saying that setting quotas of employees from minority groups could lead to resentment against them and set back the cause of anti-discrimination. What do you think of that argument?'

Tom here has ignored Bob's insult and returned to the theme. He could so easily have responded with a personal insult in return, but the argument would have gone nowhere. Of course, it may be that Bob returns with more personal insults in which case it may be best for Tom to stop the argument.

Secondly, get to know the warning signs. There are normally some physical sensations associated with getting angry: your face feels hot, your heart rate increases, you feel heightened emotions. Get to know what it feels like for you to begin to feel angry so that you can put in early preventive measures.

Also watch out for situations, words or issues that get you worked up. Some people react angrily if their authority is threatened, if their integrity is questioned, or if they feel they are being told what to do. It will be different for each person. If you know those situations you can watch out for them.

Thirdly, if you feel your pulse racing and you're beginning to get angry, keep quiet. Take a deep breath. It may be that the best thing to do is to say: 'I think we should talk about this another time.' If necessary, walk away. Go and get a drink of water or, if possible, lie down. Keep saying to yourself: 'I'm not going to get angry about this' (not too loudly though!).

Walking away might not be the ideal thing to do but it will nearly always be better than getting angry. You'll be able to address the issue better once you've calmed down. If you can't do that then count to ten slowly, or make a mental list of your friends. Do something to take your mind off angry thoughts. Plan in advance what you will think about if you feel yourself getting angry.

Fourthly, it may well help saying aloud that what the person has said has upset you. Admitting that you are upset will help you and help the other person understand the effect the conversation is having on you. You can acknowledge the other person's views quite simply:

'I realize what you have just said is your religious view, but I'm very upset by it.'

Fifthly, keep your voice quiet and well modulated. Many people who shout are unaware they're doing so. If you think you're speaking forcefully, you're probably shouting. So deliberately speak quietly.

Tip: If you're feeling at all angry you're probably becoming much more aggressive than you think.

It's very tempting to match the volume and tone of the person talking to you. If they start speaking more loudly you tend to too. Watch out for this. Don't let their anger create your anger.

Body language

There are some excellent books on body language around (try James Borg's *Body Language,* 2008, Prentice Hall). I'll outline only a few key points here. But as it is often said, 70 per cent of communication takes place through body language. Here are a few top tips:

- Don't sit or stand too close to the person you're talking to.
- Sit or stand opposite them.
- Use some, but not too much, eye contact.
- Use open body posture: don't cross your arms across your chest.

Similarly watch out for these signs from the person you're talking to:

- Are their arms crossed over their chest? If so, this suggests tension.
- Do they look shifty or uncomfortable? That could indicate they're not being entirely open.

Colourful language

Use colourful language! No, I don't mean using naughty words! I mean try to spice up your argument with some colourful words and phrases. Don't overdo it: you're not auditioning for RADA. But there are plenty of ways of making your argument attractive for listeners:

- *Use analogies*. When Microsoft was asked to bundle other companies' software with its browsers, Bill Gates said that it was like 'requiring Coke to ship two cans of Pepsi with every six-pack'. This immediately understandable analogy made the point really well. Avoid clichés. Create your own analogy. If you're trying to make the point that the other person is trying to achieve the impossible, try to think of an appropriate analogy with a well-known person: 'It's like trying to make Gordon Brown smile naturally'; 'It's like trying to teach Richard Dawkins to pray.'
- *Use 'intensifiers'*. These are words that have a strong association. Avoid words like 'very' or 'a lot' and choose words that convey a meaning dramatically. Have a look at the words used by advertisers. Bleaches don't just clean they 'destroy bacteria'; moisturisers don't just moisturise they 'soften' and 'hydrate'.
- *Choose terminology carefully*. Anyone who follows debates will become familiar with the battle over which terminology should be used. Consider, for example, the language used in the abortion debate: is it a foetus or an unborn child? Each side seeks to use their terminology because, consciously or not, it can affect the way the argument is looked at. Be careful not to adopt the terminology of the other person as it can skew the debate.

Al Smith, an American politician, when asked what his views on alcohol were, said:

'If by alcohol you mean that which is the defiler of innocence, the corrupter of chastity, the scourge of disease, the ruination of the mind and the cause of unemployment and broken families, then of course I oppose it with every resource of mind and body.

But if by alcohol you mean that spirit of fellowship; that oil of conversation which adds lilt to the lips and music to the mouth; that liquid warmth which gladdens the soul and cheers the heart; that benefit whose tax revenue has contributed countless millions into public treasuries to educate our children, to care for the blind, and treat our needy elder citizens – then with all the resources of my mind and body I favour it.'

Words too can be important in answers to arguments. Consider these two responses from a chair of a meeting:

'We have before us a carefully researched and well-argued proposal.'

This statement may make it far more likely the proposal will be accepted than:

'Right, well, that was ... err ... interesting. Is there anyone else who wants to speak in favour of it or shall we move on?'

Empower the person

The best way to argue is not to tell the other person what to do, but to get them to work it out for themselves. A person is more likely to 'own' the solution if they are part of it. This is why it can be so persuasive to give people the arguments on either side. Imagine a local meeting where Bob speaks:

> 'We're here to decide about whether to oppose the new mobile phone mast being erected. We have heard all the benefits: our phone reception will be a bit better; we will get a bit of extra money; and we will have found a use for that piece of waste land. We have heard about the disadvantages: there is a small increase in the risk our children will get cancer; the land cannot be used to build a fantastic new playground; it will destroy the beautiful views from the hills. We need to weigh these up to reach our own decision about what's best.'

You can be in little doubt where Bob's sympathies lie, but he's not telling you directly what to think. He's leaving it to you to work it out for yourself. Of course, he has set down a clear path he wants your thinking to follow.

Summary

Spend time on the presentation of your argument. Make sure you keep it simple and keep it attractive. Address not only the arguments you have in favour of your case, but also the arguments against. Use dramatic, exciting language to draw your listener into your enthusiasm for your case.

In practice

How you present an argument isn't about a new outfit or a haircut. Of course grooming matters in many situations. But in arguments, groom your words. Be clear, colourful and courageous. Be clever, concise and calm. But most of all be charming. Use humour and humility to empower the other person to see things from your point of view. Then you will win.

4
Chapter

Golden Rule 4

Listen and listen again

The goal of an argument is to explain to another person your concerns or views about something and hopefully to win them over to your way of thinking. Expressing yourself clearly is therefore crucial and we shall be talking about that later. But, if you're to persuade another person, you must listen to what they're saying.

Listen, listen, listen. It's such good advice I mention it thrice.

There are three important reasons for this:

- You will only persuade someone of something if you address the concerns they have.
- You must present your arguments in terms that the other person will find convincing.
- By keeping quiet (when listening) you're giving the other person time to present their arguments. The weakness of their view may become more apparent to them and to others, and they might very well 'shoot themselves in the foot'.

As a general rule, you should spend more time listening than talking. Aim for listening for 75 per cent of the conversation and giving your own arguments for about 25 per cent.

Tip: You want to talk with *not talk* at *the other person.*

Getting the other person to talk

Listening sounds like the easiest thing in the world, but in fact it's very difficult. The temptation is to think about what you want to say when the other person is talking. You can see this most obviously when a person interrupts another person. They are so focused on what they want to say that they're not listening.

Tip: Don't interrupt. It's rude. By interrupting you're implying that what you want to say is far more important than what the other person is saying.

Listening to someone is not just keeping quiet while they're talking. It involves trying to understand what the person is saying and why. If you don't understand, then ask for clarification. Some people will need help explaining their view. As we said earlier, some people just state their conclusions and need to be encouraged to explain their reasoning.

'That's really interesting. I've never met someone who thought that the world was flat. Why do you think that?'

Asking a question of the other person is important because it reveals to you where they're coming from and what are the foundations of their arguments. Only once you know these can you seek to challenge them.

You might find that the person doesn't know why they think what they do. You may even need to help them:

'You say you think cousins who marry are disgusting. Is that because of religious reasons? Or are you worried about any children they may have and birth defects?'

Of course some (maybe most) people haven't thought through why they have a particular view.

Address the other person's arguments

Consider this argument:

Getting it wrong

Brian: 'There's nothing for it, we'll have to sack Lucy.'

Sheila: 'But she has two young children and it will be cruel to fire her.'

Brian: She's just costing the firm too much and we need to reduce our wage bill.'

Sheila: 'But it's just coming up to Christmas, it will be hard on the children.'

Brian: 'The company is going to go bust if we don't do something to cut costs. Sacking her is the easiest way to cut costs.'

Sheila: 'You're just being cruel and heartless.'

Brian: 'We must be realistic.'

Sheila: 'You just don't understand.'

This argument is not going well. The problem, for both Sheila and Brian, is that they're not listening to what the other person is saying. Brian is not addressing Sheila's real objection to the proposed dismissal. He can make as many points as he likes about the financial wisdom of the decision, but none of those are addressing Sheila's central concern, which is about Lucy's children. Similarly, Sheila can make as many points as she wants about the children, but that's not considering the issue from Brian's perspective. It's as if they're trying to play tennis together, but each is hitting a different ball. The argument is not going to get anywhere. Brian needs to persuade Sheila that the dismissal is not going to be harsh on Lucy and her family, or to think of a way of lessening the blow. Perhaps, for example, the dismissal could be postponed until after Christmas. Sheila needs to suggest other ways of saving money if Brian is to be persuaded not to sack Lucy.

So a key part of winning an argument is listening to the statements that the other person is making and addressing them. If you don't, you'll keep making points that the other person won't agree with, and you're not addressing the reason for your disagreement.

What arguments will convince the other person?

What will mark out an excellent arguer from a good arguer is whether they can put arguments that will convince the other person. You may have a host of excellent points to support your case, but you need to choose from your arsenal the arguments that will most persuade the person you're talking to. Then you need to think about the best way of presenting those arguments, making them most attractive to the person you're arguing with. What you might think is a really good argument might not be a good argument to the person you're talking to.

Consider this discussion between Alison and Charles:

Getting it wrong

Alison: 'People on benefits are just on the fiddle and are lazy scroungers.'

Charles: 'That's not fair. My friend Mary has been trying to get a job for months. She tries really hard and it's not that easy.'

Alison: 'Well, I read this study in the newspaper last week that said that over £12 million is lost each year through benefit fraud.'

Charles: 'But Mary isn't defrauding anyone. She's a very honest person.'

Alison: 'Do you know how much of our tax goes to pay benefits? I work hard to get my salary and it just goes to pay benefits for people who don't work.'

Charles: 'But I don't mind my money going to people like Mary. She deserves it.'

This argument highlights a common problem when people argue. There are some people who focus on the big picture. They find statistics and studies very convincing. Others prefer looking at issues in relation to an individual case.

In the argument between Alison and Charles, Charles is the kind of person who finds it easier to consider issues by focusing on individual cases. So, if Alison wants to persuade him of her point of view she should give him examples of cases of people who are 'lazy scroungers'. Similarly, if Charles wants to persuade Alison of his point of view he needs to find studies or the views of experts to support it. She seems to be the kind of person who is not convinced by the stories of individuals.

In fact, most people probably find a mixture of personal stories and statistics convincing. So, especially if you're talking to a group of people or to a person you don't know very well, try to give arguments based on the broad picture as well as on an individual scenario – as in this example:

Getting it right

'We need to reorganize the layout of the office. You'll see from the plans I have given you that this will create an extra 250 square feet of space that can be used for office space and create two new desk stations. The cost per added square foot comes in at only £60. Consider, for example, Steven. He's currently squeezed into a tiny space and has to waste a lot of time walking to the other side of the office to get to his filing cabinet. Under my proposal he will be much more comfortable and not be wasting time.'

Here, the arguer has focused on the general figures and statistics, but has also given an individual example of the benefit of the proposal.

What are the other person's prejudices or assumptions?

We all come to arguments with prejudices and assumptions. Listen carefully to what the person is saying. What assumptions are they making? What kind of arguments do they seem to find convincing?

Remember that the person you're listening to may have core beliefs that you're not going to shake in the course of a short argument. You will not persuade a patriotic American that his country's foreign policy in the past two decades has been profoundly wrong. And a religious person may be more likely to be sympathetic to a religious-based argument than to one based on the assumption that there is no God.

There are less obvious points to bear in mind too. We all have views about ourselves. We have a particular image of ourselves and can get most disturbed when it's apparent that others do not see us as we see ourselves. In an argument it can be a good idea to appeal to values that a person holds dear.

> Bob: 'Sanjev, everyone knows that you're a person who keeps his word. Just the other day Barbara was saying that "with Sanjev his word is his bond". So you cannot go back on the promise you made last week.'

In this argument Bob is appealing to Sanjev's sense of identity as a person who is trustworthy. Most people care deeply about their reputations and how they are thought of by others. Appealing to a person's core values and seeking to connect your arguments with those will be persuasive.

> **Useful example**
>
> 'If you do this people will think you're dishonest and manipulative. Do you want to be seen as that kind of person?'

Who does the person respect?

Finding out who the person you are arguing with respects or trusts is important. Imagine you know the person you're arguing with is a passionate supporter of Barack Obama. It will be a powerful tool if you can point out that their view goes against

Barack Obama's. At the very least, you should be able to say to them: 'Look if Barack Obama disagrees with you, don't you think that at least you need to think about the issue carefully?'

This is important too when considering which statistics you should use. If you know the person is a keen supporter of a particular children's charity, say, then see if you can find a study by them supporting your conclusion. At the very least avoid statistics from organizations that the person you're arguing with opposes. A militant atheist is not going to be convinced by a report on the power of prayer prepared by the Church of England. They would be more convinced if you could find a report from an atheist concluding that prayer can do some good.

Find common ground

A key to success in an argument is finding some common ground. Are there facts that you can agree on? Until there are some agreed facts it's hard to proceed. Consider this argument between parents.

Getting it right

Mum: 'We must stop Tom watching *Dr Who*. He's watching too much TV.'

Dad: 'OK, but he loves *Dr Who*. It would be really hard to stop him.'

Mum: 'Yes, but are we agreed Tom is watching too much TV?'

Dad: 'Yes, I agree.'

Mum: 'And he has watched two hours today already?'

Dad: 'That's true.'

Mum: 'So he shouldn't watch any more.'

Dad: 'Good point, let's record it and he can watch it tomorrow.'

Mum: 'Good solution. Shall we say there is an absolute rule that he can watch no more than two hours of TV a day?'

Dad: 'Yes, that's a good rule.'

This argument worked well. It could easily have gone wrong. Mum did well to move to establish some facts they could agree on. Dad, once he saw the key facts, agreed and they were able to find a solution.

There's another lesson from this argument. The use of pronouns can be important. Talking of 'we' can bring in the other person, and is a useful way to highlight what you agree on.

Useful examples

'Let's try and establish what we agree on.'

'Could you explain that to me again? I'm having difficulty understanding your point.'

'We need to find a solution to this that we can both live with.'

Link up with a person's positives. Where possible find areas of agreement:

'I agree you've made some great points in that presentation. However, we do need to weigh up the disadvantages with the benefits.'

Everyone likes compliments and, even though it sounds old-fashioned, flattery. Just because you're in an argument with someone doesn't mean you can't be nice to them!

But what if you can't agree on the facts?

Sometimes it's not possible to agree on the facts. In that case it may be that the argument is not going to go anywhere. In the preceding argument involving the parents, if they couldn't have agreed on whether Tom had or had not watched television earlier in the day, it would have been difficult to resolve the argument.

Sometimes it's useful to proceed in a discussion on the basis that a particular fact is true. For example, you might say: 'Look, let's assume that X is true, if so I agree with you.' You make it clear that you do not necessarily agree that X is true, and indeed if X turns out not to be true you will not agree.

This is particularly helpful if you think you have a strong case, even if your claim is wrong.

> Bob: 'You think we should fire Lisa because she has lied to us. Now, we disagree on whether or not she has lied. But for the moment, let's assume that she has lied. Even then, I still think we shouldn't fire her. She's never lied before and she's a hard-working employee.'

Unless Bob makes this tactical move the argument may be stonewalled on the debate over whether or not there was a lie. However, if Bob succeeds in his argument that Lisa should be kept on whether or not she lied, then it matters much less whether or not there was a lie.

A similar tactic can be used to find a 'contingent solution' to the argument:

> Wu: 'Well, I know we're disagreeing about how much this project is going to cost, but what about this as a solution? We'll get the accounts department to cost this project. If they decide that it costs under £30K we'll go ahead, but if it's more we will not.'

In a case like this, where the facts are unknown or disputed, there's not much point carrying on the argument until the facts are known. It's better either to stop the argument until the facts are known, or to reach an agreement that will depend on the facts once they are known.

Summary

So, in all sorts of ways, listening has its advantages. You learn the other person's counter-arguments, which you can then address. You find out their perspective, and then have insight into which approach might best convince them of your point of view. And who knows, when you give them free rein to spout off, they might very well dig themselves a hole they can't get out of. So listen, listen, listen. I mentioned it thrice, it's not such a price.

In practice

When listening, be careful not to daydream about what you're going to say next. Practise attentive listening, where you are digesting exactly what that other person is saying. In doing so, you will add depth to your own argument, and be able to build common ground from which to move forward.

Chapter

Golden Rule 5

Excel at responding to arguments

As I've said several times already, being a good arguer means not only making the points that you want to make, but also responding to the points that other people have made. The best form of argument involves putting forward your best arguments and seeking to counter the other person's.

There are three ways of responding to an argument:

- Challenging the facts upon which the other person is relying.
- Challenging the conclusions they are reaching.
- Accepting the point they have made, but arguing that there are other points that outweigh what has been said.

This will be clearer if we look at some examples.

All English people dress badly. The Queen is English. Therefore the Queen dresses badly.

Here there are two premises: that all English people dress badly and that the Queen is English. From this there is the conclusion that the Queen dresses badly. The logic here is flawless, but if you wanted to challenge the argument you can challenge the first premise – is it correct that all English people dress badly? Can you think of an English person who dresses well? (Probably not!) You might challenge the other premise – the Queen is English – but that seems harder to challenge. As this example shows, sometimes you cannot fault the logic of the argument, but you can challenge the accuracy of the statements (premises) used as the basis of the argument.

The Pope is Catholic. The Pope opposes abortion. All Catholics oppose abortion.

For this argument there are two premises: that the Pope is Catholic and that the Pope opposes abortion will be readily agreed by most people. But here the conclusion does not follow from the premises. Just because one person who belongs to a group has a view does not mean everyone has that view.

This is another bad argument:

A banana is a fruit. A banana is yellow. All fruit is yellow.

In both these arguments, dodgy conclusions are reached from indisputable facts. So conclusions do not always flow from the facts. Challenging a conclusion someone has made is the second excellent way of responding to an argument.

The third form of challenging an argument accepts the premise and conclusion, but asserts that the argument ignores other factors.

Walking to school is healthy. We want to be healthy. We should walk to school.

Assuming for the moment the premises and logic of this argument are correct, as they probably are, it's not the end of the argument. We might want to be healthy, but there are other things that we want (e.g. getting to school on time, arriving in a cheerful mood) that need to be weighed against this argument. Also, there are different ways of being healthier that might fit into the day more easily. So while walking to school is laudable, the conclusion reached could be challenged by raising other factors that might pertain to this particular situation.

Let's look at these different ways of responding to arguments in more detail.

Challenging the facts

Imagine Bob says this:

'Average temperatures in the UK are falling, not rising, and therefore talk of global warming is nonsense.'

One way a person concerned about global warming could respond is by challenging the factual basis of the argument. They might be able to produce a survey that showed that in fact average temperatures in the UK are rising.

As we have already seen, statistics and studies can be misleading. We saw in Golden Rule 1 how easily statistics can be misused. Remember some of the key things to ask about statistics:

- Who did the study? Was it independent?
- How big was the sample? Was it representative?
- Exactly what does the study show?

It is easy to be taken in by 'science'. Beware of someone trying to bamboozle you with long words. Madsen Pirie, a leading expert on logic, gives an example of how something very simple can be made very complex:

The small, domesticated carnivorous quadruped positioned itself in sedentary mode in superior relationship to the coarse-textured rush-woven horizontal surface fabric.

Or, in more simple words: the cat sat on the mat!

It's a common trick in arguing to make something sound very complicated. You are not then able to understand it and argue against it. Indeed, the trick is to make you believe that you're not clever enough to understand their point and therefore agree to whatever they say. This is particularly common in academia. My experience is that, in fact, really clever people are able to explain their ideas in a very straightforward way. It's those who are less clever who feel the need to dress up their ideas with long words or complex ideas. You should never feel embarrassed about asking someone to explain what they're saying in terms a reasonably intelligent person can understand. If they can't, the fault lies with *them*, not with you!

Tip: Some people seem to live by the principle 'Never use a four-letter word when a fourteen-letter word will do'. Beware of them!

Relying on experts can be a problem if you assume that as they're expert in one area, they're expert in all. Of course we can see that they are knowledgeable in a particular field, but in many other areas you could be far better versed. Certainly don't listen to an academic expert on the topic of cars (an example of my extreme bias in making this presumption). I know many professors who are world experts in their fields, but there are few of them whose views on which is the best barber in Oxford I would accept; indeed, their appearance tells me more about which are the worst!

It is remarkable how the press, in particular, is willing to listen to the views of an expert on one area, and assume they are an expert in all areas. Film stars, for example, often weigh in on complex political matters and are asked for their political insight, information that would better come from an analyst in the field.

It's always wise to ask experts what others in their field think. I'm always careful in my lectures to make it clear what is known as a fact; what most academics think on an issue; what we don't know for sure; and what I think (which is not always the same). So I will commonly say, 'The general view amongst lawyers is that the courts would interpret the law in this way, but my own view is that the courts should interpret it in another way. Let me give you the arguments on either side ...' A good expert should easily be able to tell you not only their view, but the views of others in their area.

So never be nervous about challenging the facts. If you think they might be wrong, then others would probably think the same. In challenging the facts, you might get an answer that actually substantiates them. This would mean the argument could move forward towards a resolution. But the facts might very well not stand up to query. You won't know unless you challenge them. If there is any doubt at all in your mind, verify the facts before you move on to the next step.

Challenging the conclusions

Remember Bob's argument:

'Average temperatures in the UK are falling, not rising, and therefore talk of global warming is nonsense.'

You might accept Bob's premise (that average temperatures in the UK are falling) but argue that his conclusion is mistaken. For example,

'Bob, you're right that average temperatures in the UK are falling, but in many other parts of the world temperatures are rising. It may be getting colder here, but that doesn't mean that the world overall is not getting warmer.'

Challenging the conclusion is useful when the facts themselves are not in dispute but you think that the logic leading to the conclusion is flawed. One particular aspect is when individual facts lead to generalized conclusions (as in the earlier example of the Pope being Catholic and opposing abortion). Every now and then a case is reported in the newspapers that produces what appears to be an unfair result and naturally the cry goes up, 'the law needs to be changed'. But we need to be careful. Just because the current law has produced an unfair result doesn't mean that an amended law will not also produce an unfair result. Indeed, it may be that whatever the law says there are going to be some cases where there is an unfairness.

So, if you want to challenge someone's conclusion you will want to argue that their conclusion does not follow from their premise. There may be other conclusions that could be reached. Ask the person you are arguing with why it is they reach their conclusion, rather than an alternative. Consider this example:

Getting it right

Bob: 'Your child is constantly yawning in my class. He clearly needs more sleep.'

Mary: 'Well, he could be yawning because he's bored, rather than tired. No other teachers report his yawning in their classes. If he was tired you would expect him to be yawning in all his classes.'

In this example Mary has effectively shown that several conclusions could flow from Bob's premise (Mary's son is yawning in class). Bob has concluded that it's because the son is tired. However, as Mary has pointed out there are a number of other conclusions that could result from the premise (the son may be bored or may be tired). Mary has gone on to introduce evidence to suggest that Bob's conclusion is less likely to be correct than Mary's.

Challenging with other factors

It's very helpful in an argument to be clear about whether you're rejecting the worth of the other person's argument or whether you're suggesting that their point is outweighed by other factors. Consider, for example, two people arguing over whether having a new supermarket in their town is going to improve the town's life. The pro-supermarket person might say:

'This is excellent news, because it will mean we'll have a far wider range of goods available in the town than we do at the moment.'

The other person has two choices. One would be to reject the argument:

‘I don’t think that’s right because the opening of the supermarket will force many of our specialist shops to close and we will end up with a narrower choice.’

Or he can accept the point but draw his opponent’s attention to other factors that need to be taken into the balance:

‘You’re quite right that there will be a wider range of goods. But there will be a lot more traffic in the town. We need to decide which is more important: having a wide range of goods or having a peaceful town.’

Try to be as clear as possible whether you are agreeing with the other person’s point or not. Otherwise you will find they are likely to state their point again.

Compare the following two examples of the same argument.

Getting it wrong

Max: ‘We should go to my Mum’s for Christmas because she’ll be very sad if we don’t.’

Susan: ‘We’ll have much more fun if we go to my sister Beth’s.’

Max: ‘I don’t think you’re quite seeing it from Mum’s point of view.’

Susan: ‘We need to think about what’s best for us.’

Getting it right

Max: 'We should go to my Mum's for Christmas because she'll be very sad if we don't.'

Susan: 'That's a good point. She loves it when we visit. But we have gone to her for the past three years, and it would be such fun if we went to my sister Beth's.'

Max: 'You're right, we always have a great time at Beth's, she's a great hostess. Is there any way we can get to see them both over Christmas?'

The second argument is much better because both are acknowledging that the other has made a good point and making it clear they accept the strength of what is said.

Another common technique in argument is to form an alliance with the listener by emphasizing common ground. Consider this point:

'We all want to make the decision that is best for the company and we must therefore accept this plan.'

The message is given that those who don't support the plan are not seeking the best for the company. Similarly this argument:

'We're all Muslims in this room and therefore we must combat immorality and oppose this plan.'

Again the listener is given the impression that support for the plan will be being disloyal to Islam. The impression may, of course, be incorrect, but it is a technique that makes the argument sound more attractive.

Summary

So remember that in working through an argument, you can accept the opponent's facts and initial conclusions, but still find points that outweigh the argument and make your conclusion attractive. By presenting other ways of looking at the situation, or bringing in other material that might not have been considered, you can get the argument to go your way. Think outside the box, and don't limit yourself to a prescribed way of looking at a situation. Imaginative arguing can win the day, as you find points that outweigh your opponent's.

In practice

Listen very carefully to the person you're arguing with. Check whether they have understood your points. What issues are really troubling them? What kinds of arguments will be most persuasive for them?

Chapter
6

Golden Rule 6

Watch out for crafty tricks

There are some nasty tricks that people can play when making arguments. Here are some. Watch out for your opponent using them against you.

> **In arguments you need to be subtle, watchful, alert and curious.**

Attacking the person

Lady Astor to Churchill: ‘Winston, if you were my husband I would flavour your coffee with poison.’

Churchill: ‘Madam, if I were your husband, I should drink it.’

An all too common way of arguing is to avoid the argument and to attack the person. For those who like clever-sounding Latin, this is sometimes called an *ad hominem* (to the man) argument. Consider this:

> **Getting it wrong**
>
> Alf: 'I think we need to take into account ethical values when we develop our investment policy.'
>
> Susan: 'You're a right one to talk about ethics, given your personal life.'

Susan's response is unlikely to be productive. It will certainly inflame Alf and is unlikely to be attractive to those listening. Indeed, it may even cause listeners to be embarrassed for Alf and support him when they would not have done so before. Alf's best response would be to try to focus again on the issue:

Alf:	'Well, we can talk about my personal life another time if you like. But we're discussing investment policies here.'

Of course there might be times when a personal response is appropriate. It might be that you're discussing a person's qualities for a job or issues of personal morality. However, generally you should be very cautious about making an attack on a person, rather than the argument. It rarely gets you anywhere.

Tip: Avoid phrases like:

- *'You're just impossible.'*
- *'You just think you're so clever.'*
- *'There's no point arguing with you.'*

Beware of causation

A common error with statistics and surveys is to assume the cause behind a particular fact. For example, it's sometimes claimed that people should marry because the unmarried suffer higher rates of poverty. The suggestion is that unmarried people would be richer if they married, but that assumes that being unmarried causes poverty. That assumption cannot be made. It may be, instead, that poor people are less likely to marry. Similarly, it's true that those on diets are more likely to be obese than those not on a diet. But that does not mean that being on a diet makes you obese! Mistakes of this kind regularly arise.

- 'Whenever ice cream sales rise, so do shark attacks.' (So does eating ice cream make you delicious?)
- 'As more economists are recruited to the Treasury, inflation rises.' (Do economists cause inflation?)
- 'As vocabulary increases in infancy, so does appetite.' (Does talking make you more hungry?)

If there is some evidence that two facts are linked, do not assume that one causes the other. As these examples show, making that assumption can lead you into error. In fact, finding cause can be very difficult. Much research has been carried out into what makes people thin or obese, or what makes people smoke. The answer, unsurprisingly, is a whole host of factors. And be alert to your opponent assuming that one fact causes the other. It's a certain way of finding a hole in their argument.

It's easy to slip into an error of assuming that because a common cause of B is A, then if B occurred so did A. Consider this argument:

When Bob gets drunk he does not come into work. Bob has not come into work, therefore he is drunk.

This, of course, does not necessarily follow. Bob may not be in work for any number of other reasons. Logicians call this the danger of affirming the consequent. Of course, if Bob regularly misses work due to drunkenness it becomes more likely that this is the reason for his absence. But we must not assume it is necessarily so.

So, when arguing, watch out for your opponent arguing that something is so because they have made an assumption about what happened before. Get them to agree to prove what has actually happened before you accept that their conclusion is correct.

The dangers of negatives

There are dangers in arguments based on what statistics have *not* proved. For example, consider this argument:

Many millions of pounds have been spent on trying to find extra-terrestrials and none have been found, therefore they don't exist.

Of course, the fact that studies have not proved something does not mean that the thing is not true, or indeed that the thing *is* true. Many great minds have pondered the question of whether there is a God, with differing conclusions. But just because no one has been able to prove that there is a God does not mean there cannot be. Just as the fact no one has been able to prove there is no God does not prove that there is.

A useful point to bear in mind here is that where there is a lack of evidence we tend to rely on what we expect normally to occur. If I were to tell you that I had met the Queen yesterday and produced a dated photograph of me and the Queen standing side by side, that may be enough to convince you that what I said was probably true. If, however, I was to say I met a Martian yesterday, and produced a photograph, that would probably not convince you. Indeed, I would need to produce an enormous amount of evidence to persuade you. That's because it is not implausible that I met the Queen, but most people start off with a heavy assumption that Martians do not exist.

Similarly, at work, someone might say that the last deal your company did with X Co did not work well. That is interesting, but it doesn't mean that all deals with X Co will be unsuccessful.

The dangers of 'illicit process'

A common error in an argument is 'illicit process'. It is best demonstrated by an example:

All vegetarians disapprove of eating meat. All vegetarians are worried about global warming. Therefore all those worried about global warming are vegetarian.

That, of course, does not follow. Just because some people worried about global warming are vegetarian does not mean all who worry about it are. Here there has been an 'illicit' process from one fact to another. Don't be taken in by arguments of this kind. Test carefully whether the arguer is assuming that all people of a particular kind are the same.

The false choice

Using a false choice is a common device in arguing. It presents the listener with only two alternatives. George W. Bush became famous for this when he spoke of the war on terror:

'You are either for us or against us.'

This gives you only two options: to agree or disagree. Of course, you may want to agree in part, or be neither for nor against the proposal. But the rhetorical device closes those options for the listener.

Parents soon become experts at this:

'You can either eat your greens or go straight to bed.'

There are, in reality, many other options for the child but the parent has presented the child with just two.

As both the examples show, 'bifurcation', to give it its technical name, is a particularly popular form of argument where one of the alternatives is seen as highly unpleasant. The child doesn't want to go to bed and so takes the option of eating the greens. Those listening to George W. Bush who did not want to side with terrorists were left with the option of siding with George W. – even if that meant siding with a man who said:

"I'm telling you there's an enemy that would like to attack America, Americans, again. There just is. That's the reality of the world. And I wish him all the very best." George W. Bush

Sometimes a good arguer can turn the bifurcation argument on its head. Consider this argument:

'If we build a new railway station here, either it will be empty and a waste of money or it will be full and the nearby roads will not be able to cope.'

One reply would be:

‘Well, if we build a new railway station here, either it will be empty and the nearby roads will be able to deal with the traffic or it will be full and it will have been a financial success.’

Further, the false-choice argument is an example of a particular use of the burden of proof, the best known example of which is Pascal's wager. Blaise Pascal was a renowned mathematician and philosopher who lived in the seventeenth century. He produced what he thought was a convincing argument for why everyone should believe in God. It went like this. Either there is a God or there is not. If there is a God and you don't believe in him you may go to hell. If there is no God and you believe in him, you might have less fun in life but you don't lose out much. Therefore it's better to believe in God. Another version of this argument is sometimes heard in the climate change debate:

‘If climate change is man-made and we cut CO_2 emissions we might save the planet.’

‘If climate change is not man-made and we cut CO_2 emissions we won't lose anything, except perhaps suffer some economic harm.’

The choice again is presented in such a way that there are two alternatives: one has a potentially terrible loss (eternal damnation, loss of the planet) and only a small gain (less fun in life, some economic harm); the other has no terrible loss, but a huge potential gain (eternal life, saving the planet). So presented, the argument seems compelling to choose belief in God and cutting CO_2 emissions.

In many ways these are convincing arguments but it can be difficult to persuade people that they should avoid the awful possibilities mentioned, even at the cost of some minor inconvenience.

The best counter to such arguments is to suggest that it is not as straightforward as the two possibilities mentioned. In relation to God there is the question of which god to believe in. There are so many gods and if you chose the wrong one you might still end up in damnation. Similarly in relation to CO_2 emissions, the argument hides the option of slightly cutting CO_2 emissions with less economic loss.

An alternative is to explore the likelihood of the events being true. If you think it's just possible, but very unlikely, there is a God you might think the risk of eternal damnation is worth running in order to enjoy the 'pleasures' of this world. If, however, you think it very likely that climate change is man-made, the argument presented above may become overwhelming.

When faced with a false-choice argument, first recognize it for what it is: a false choice. You can explore the likelihood of the events being true, as well as look for ways in which the arguments are not as straightforward as they appear. By this means you can bring meaning to the discussion and depth to the argument.

Generalizations

It's always tempting when having an argument to make generalizations:

'You never do the washing up.'

'Politicians don't understand what it is to be poor.'

These kinds of comments are really asking for trouble. It's nearly always possible to think of exceptions. The person you're arguing with can easily come up with a counter-example ('Well, I did the washing up last Sunday'). Your point is then weakened and indeed you're even open to the charge that you're exaggerating or lying. In the examples just given, if you wished to make the point you could do so in a way that would be more attractive:

'You don't do the washing up very often.'

'Many politicians don't understand what it is to be poor.'

Of course these statements may still be untrue, but they are much more likely to be true than the generalizations mentioned earlier.

Do be careful of the use of individual cases. Consider this:

'Everyone is so rude these days. Just yesterday a person knocked into me and didn't even apologize, they just walked on.'

It's easy in this case to seek to challenge the example raised. You may offer possible reasons for what happened that would explain the apparent rudeness. Maybe the person who knocked into her didn't speak English and so couldn't apologize. However, usually a better way to respond is to produce examples of your own where people were not rude. In fact, if you're trying to show that a generalization is untrue you are on a stronger wicket than a person seeking to show it's not. So to dispute the claim:

'All English people are good at queuing'

all you need to do is to show a single example where that was not true. However, typically, to support such a claim a single case (which could very well be a one-off) is used.

Like cases

A key principle of logic is that if two cases are the same then a reason must be provided for not treating them in the same way. Hence a popular tool in arguing is by proposing the analogous situation:

'You say that we should stop people smoking because it makes people ill. Do you support stopping people eating fatty foods?'

This is a perfectly fair argument. It's helpful because it will elucidate why the person thinks the way they do. It might highlight the fact that their opinion is based on prejudice. If the smoking-banner were to reply:

'Well, I really enjoy fast food so I don't want to ban that'

then they lay themselves open to a charge that they want to ban the vices that other people have, but not their own! They need to produce a good reason to distinguish the cases or agree that both are the same. So instead they could argue:

'Well, the vast majority of smokers die from smoking-related illnesses, but few eaters of unhealthy food die from their diets.'

Of course it would be necessary to back up the factual claims made there. Alternatively they could argue:

'Yes, quite right. As citizens we owe each other a duty to keep healthy. All clearly unhealthy behaviours should be banned, be that smoking or eating unhealthy food.'

For the 'like cases' tactic to work, you might need to apply your position to what might seem an odd view, but not necessarily wrong. Say, for example, you are committed to opposing sex discrimination. You're then asked, 'Well, do you think women should be able to apply to be boxers?' The answer should be, 'Yes, why not?' If you are committed to your principle then unless you have a good reason you should stick to it even if the consequences seem odd. But beware, it may be that you are being tricked. 'Is a film director entitled to refuse a woman the role of Winston Churchill?' The answer may be: 'Yes, as long as he is not refusing her the role because she is a woman, that's permissible. If other candidates look more like Churchill they may be better suited to the role.'

Red herrings

These are important. They involve introducing completely irrelevant material.

> **Getting it wrong**
>
> Sami: 'How dare you forget my birthday!'
>
> Raj: 'You know, you're so handsome when you get upset.'

Quite clearly here Raj is aware that he has no excuse for forgetting the birthday, and is trying to introduce a new topic of conversation on which he is far more comfortable: the handsomeness of his beau.

This is, in fact, a common way of dealing with a brewing argument in social situations.

'Well, this is a really interesting discussion, but I'm afraid I must get ready to go out. Did I tell you we were going to see this new film?'

Often both parties, if they are friends, are happy to avoid the controversial issue and discuss the more pleasant topic of the film. Normally, introducing a red herring is an invitation to abandon the argument and discuss something different. You'll need to decide whether or not to accept the invitation.

Some red herrings are deliberate attempts to muddle you.

> **Getting it wrong**
>
> Alf: 'Abortion is murder and should be outlawed.'
>
> Brian: 'That's a bit harsh; why do you think it murder?'
>
> Alf: 'Well, it's killing a child.'

Brian: 'But it's not really a child, it hasn't got feelings or thoughts.'

Alf: 'Well, Brian you're not a parent and I don't think you understand about children.'

Alf has deliberately set off on a change of focus. Brian needs to bring it back to the subject.

But not all red herrings are appreciated. Let's go back to the man who forgot the birthday. I'm sure we have all wanted to upbraid someone about an issue only to find they keep changing the subject we're trying to address. It can be infuriating! Both parties need to beware here. The red herring can be a clear sign that you don't want the argument now, but pay attention to whether the other party takes the bait.

Getting it right

Sami: 'How dare you forget my birthday!'

Raj: 'You know, you're so handsome when you get upset.'

Sami: 'That's kind of you to say, but I want to talk about why you forgot my birthday.'

Using a red herring can be dangerous too. Is this an argument that needs to be had? It may be that if the issue is not resolved now it will always be sidelined. Is this perhaps the right time and place for this argument? Is it an argument over a topic that might actually be productive? At least recognizing a red herring will give you a choice on how to proceed. And it's a useful tool to employ yourself if you ever get stuck at the watercooler discussing your salary!

Circular argument

This is another deceptive type of argument to watch out for. It uses two unproven facts to bolster each other and give each credibility. Consider this:

‘God exists because the Bible tells us so. We can trust the Bible because it is the word of God.’

All of this may be true, but this argument is not a good one! Arguments from logic require us to start with a fact that is true and reason from this. The difficulty in this example is that A is only true if B is true, and B is only true if A is true. Another example of a circular argument is this:

‘I'm better than you at arguing. You always end up agreeing I'm right. You should accept I'm the better arguer.’

Concealed questions

A clever technique that is sometimes used is to ask a question that contains a hidden fact. In answering the question the person is thereby assumed to accept the fact. The best-known example is:

‘Have you stopped beating your wife?’

Whether the man answers yes or no he is admitting that he has beaten or is beating his wife. More subtle forms would be:

‘Has your unethical approach affected your profits?’

This can be a crafty device to get the hidden fact accepted. Lawyers in courtrooms use this technique a lot. The question:

‘Who was the woman you were with on the night in question?’

assumes there was a woman and can trick the witness into accepting that fact if they're not very careful. If the witness replies 'I don't want to say, it's private', the witness has admitted he was with someone.

This is a clever trick to learn if you want to get other facts from your opponent to further your argument. So if you're wondering whether your wife is really going to her exercise class at the gym, or whether she might be having an affair with Brian, you could ask 'How is Brian these days?'

Literalism

One of the most annoying kinds of arguments can be those with people who rely on literalism. Literalism is the joy of lawyers and insurance companies. They base their argument on the literal meaning of the words they have used, rather than how those words would be understood by ordinary people. Hence you get the argument:

> 'We told you we would supply you with a new car, but we did not claim it would work.'

You can spot a literalist by some tell-tale words: 'Let's look at my exact words ...' or 'All I said was ...' The annoying thing is that their arguments can often carry much weight in a court of law. In a contract dispute they will only be bound by what they promised to do. Indeed, if you're in a dispute over whether you have broken your word it's well worth thinking carefully about what you said you would do.

So what can you say to your literalist? One response is to see whether you can turn the tables on them. Maybe you agreed to pay them, but never said when you would. This way you can turn the tables on them and say: 'If you are going to take your obligations literally, so will I.' That may lead to an agreement to read the contract in a sensible way.

Alternatively, you could ask what they meant people to think when they said what they said. A good response to a literalist is to suggest that had they wanted to say what they claimed, they could have done so clearly. Consider this:

Getting it right

Shazia: 'All I said I would do is that I would give you a refund. I did not say it would be a full refund.'

Mary: 'But anyone who is told they would get a refund would think it would be a full refund.'

Shazia: 'Ah, but you must listen to what I said.'

Mary: 'I did. If you had wanted to be clear you could have said it would only be a partial refund. By not making that clear I relied on the normal meaning of the word refund.'

Mary is making some good points here. She may not persuade Shazia, but she's making her argument well.

Sometimes it is best to give up arguing with literalists.

Hostile association

This form of argument is to cast doubt on a viewpoint because it is one held by disreputable people. For example:

'You don't want to be vegetarian. Hitler was one of them.'

Here you're suggesting that supporters of vegetarianism are associated with Hitler. Of course, that's completely unfair. Wicked people just occasionally have non-offensive views. It's really quite difficult to be wrong all of the time about everything!

Sometimes 'hostile association' is more subtle and relies on a hearer's prejudice:

'A right-wing think tank has suggested lowering taxes, but ...'

Such a speaker is relying on you immediately dismissing any idea that has come from a right-wing think tank. Similarly if, let's say, the accounts department in your firm is particularly unpopular you could argue:

'Now, this proposal is very popular with the accounts department, but ...'

'Begging the question'

'Ah, but you're begging the question,' people often complain. The term 'begging the question' (officially known as *petitio principii*) is commonly used, but not always properly. It is correctly used where a person puts forward an argument that is in fact no more than a reworking of their conclusion. So, rather than relying on a premise to argue to a conclusion, they use a conclusion to argue a reworded conclusion.

'Abortion is murder because it involves the killing of an innocent child.'

Well, 'the killing of an innocent child' *is* murder and so in effect all that is being done here is to restate the conclusion, but create the impression that an argument has been used. You can normally spot such an argument if you think, 'Well, anyone who believed your first point would agree with the second'.

Getting it wrong

'This deal will make an excellent profit. We will therefore quickly recoup our losses. Those who think the deal involves taking on dangerous debts are therefore mistaken.'

In that argument the conclusion that there are no dangers with the deal is only true if it's true that the deal will make an excellent profit. You can often spot a begging-the-question argument because it's one that no one could disagree with, if only the fact it started with is correct.

Slippery slopes

This is a common device that arises in arguments. It centres on the issue: where do we draw the line? Consider, for example, an argument about whether the NHS should deny treatment to those who have diseases caused by smoking. An opponent may argue:

'Where next? Will we deny treatment to those who are overweight, to those who do not exercise enough? You will end up with the NHS only offering treatment to superfit, ultra-virtuous athletes.'

In a 'slippery-slope' argument, the arguer seeks to show that there is no logical place to draw the dividing line, that once one exception is accepted then a line cannot be drawn anywhere sensibly. You're therefore driven to accepting an absurd conclusion. Because you don't want to reach the absurd conclusion you decide it would be better not to take even one step down the slippery slope. For example, schools and colleges often have absolute policies on matters such as uniforms, for fear that once one exception is granted there will be a flood of requests for more exceptions.

In a slippery-slope argument, you make the point that once we allow A, we must also allow B, C, D and E, as there is no good reason for distinguishing them from A. Your argument is that having to accept D or E will be disastrous. We must, therefore, not allow even the exception of A, however innocuous it might look on its own.

Responding to a slippery-slope argument can be difficult. There are two ways this can be attempted.

- *Deny that the slope is slippery.* One response is to suggest that the place where you had drawn the line is a justifiable one and there is no reason why you need to accept that other scenarios would follow.

'I think we can allow an exception to the uniform code for this student because this involves a religious belief. We can explain that only exceptions based on religious belief will be allowed, and there will not be many of those.'

- *You could argue that the slope is slippery everywhere.* The argument here is that there is no sensible reason to draw the line where you do but it has to be drawn somewhere. Take, for example, the fact that to buy alcohol you need to be 18 in the UK. Now, it's easy to argue that this is an arbitrary line. Nothing magical happens on the night of an eighteenth birthday. But this point can be made at whatever age is chosen. It can always be argued that the child does not magically become more mature in the course of a few hours the day before the relevant birthday.

This is true in lots of areas of life. Take speeding. Is it really that much more dangerous to drive at 31 mph than 29 mph? Probably not, but driving at the first speed can cause you to end up with a ticket, the latter not. So the first question to ask is whether there needs to be a line drawn somewhere. Well, assuming we don't want 7-year-olds buying beer and we don't want roads without speed limits, we need to draw the line somewhere. Having reached that conclusion, we need to accept that wherever the line is drawn there will be cases either side of the line where it seems arbitrary. The next question is whether the place the line is drawn is a reasonably good one. So, regarding the age to buy alcohol, we are confident that, generally, under 18-year-olds lack the maturity, needed to make the decision to buy alcohol, while over 18-year-olds do possess it. If that is correct, there is a strong argument for saying, yes the line drawn is an arbitrary one, but we have to draw a line somewhere, and this place is the best place to draw it.

What if?

A common tactic in argument is to produce an absurd scenario that will produce disaster.

‘Bob has suggested that we relocate to Milton Keynes, but what will happen if there is a national rail strike?’

Or more dramatically:

‘That financial plan looks very sensible, but what would happen if there is a stock market crash?’

This form of argument is common. Its essence is very sensible: it can be used to point out the dangers of a proposed course of action. However, the argument should be treated with care. Virtually any idea could be opposed on the basis that one can imagine a scenario where it would be foolish. ‘But what if ..?’ can always be asked. For example:

‘We should not buy Christmas presents this year, because Martians may land tomorrow and take over the world.’

A good use of ‘what if?’ is to show that not only are there potentially disastrous consequences, but that those are realistic. If you are the opponent of a ‘what if’ argument, your case will be even stronger if you can show that there are alternatives that are just as good, but that do not carry with them the suggested disadvantages. Also, many ‘what if’ arguments can be defeated by saying that the dreaded scenario will be a problem whatever the situation is. If Martians land and take over the world, the fact you have bought Christmas presents may be the least of your worries!

Straw men

The battlefields of arguments are littered with the bodies of slain straw men. Straw women seem rarely to face attack. It can be a powerful rhetorical tool to pick on a particularly weak argument that could be used by the other side and ridicule it. Consider these two examples:

'I read yesterday an article from my opponent's party arguing we should raise taxes so that we can spend money on improving Buckingham Palace. Well, I think the Queen can well afford to look after her own building rather than calling on the downtrodden taxpayers of this country. So I say: No More Taxes.'

'The best argument that people come up with for playing Peter Crouch as a striker in the England football team is that he's tall. However, what we want in strikers is that they can score goals, not that they're tall.'

Both of these arguments are designed to make their opponents look foolish. However, both are based on the false assumption that the only arguments opponents might make are those mentioned. Of course, much better arguments could be used for raising taxes or picking Peter Crouch than those mentioned.

A version of the straw-man argument is to characterize your opponent's arguments in the most extreme way possible:

'I hate green politicians; they would close down every factory in the country if they had a chance.'

'Those who support cuts in the defence budget want to leave our country open to be invaded.'

The best way to defeat the straw-man argument is clearly to dissociate yourself from the ridiculous argument:

'I agree with my opponent that raising taxes to improve Buckingham Palace would be absurd. But I can think of much better things to use tax money on than that. What about improving our hospitals. Would you agree to raising taxes to spend more on hopsitals?'

The dangers of the two-wrongs argument

You are likely to have come across the two-wrongs argument: 'Bribery is OK because everyone's doing it.' 'If we don't sell arms to this unpleasant regime someone else will.' But just because someone else is doing something wrong does not mean that it becomes permissible to do it. We would think it a very bad argument if a paedophile were to say: 'It didn't matter that I abused this child, because if I had not someone else would have.'

So beware of those who use the two-wrongs argument, and be careful of using it yourself. It is never a justification for any position.

The power of silence

It's important to realize that in an argument silence is an option. Indeed, it can be an important tool. I'm sure we have all been at meetings where the longer a person has spoken the less we are convinced by their position. Especially in a meeting, it can be more worthwhile letting someone continue to make their case very badly than trying to intervene.

Silence is, of course, key in order to avoid an argument altogether. Remember Golden Rule 2: there is a time and place for every argument. If you're not sure whether this is the time or place, it's probably best simply to remain silent. Silence is in its nature equivocal; you should not be taken to disagree or agree with what is being said. If you are pushed to respond you can simply say: 'I'm not ready to discuss this right now.'

Silence may also be an appropriate response where you feel that the person you are arguing with has made a very good point to which you don't have a ready response. Being silent may encourage them to make another point to which you may have a better answer.

"Silence is one of the hardest arguments to refute." Josh Billings

Feeling stuck?

Sometimes in an argument you may feel you don't know what to say. It may be best to suggest that you continue the argument another time so that you can clear your head. If not, it's useful to have a stock of phrases you can use:

> **Useful examples**
>
> 'Could you explain that in non-technical language?'
>
> 'What are your parameters?'
>
> 'Are you just begging the question?'

While the other person is dealing with your questions you will have some time to think about what you want to say.

Summary

Watch out for arguments that at first seem convincing, but on closer analysis are not proper arguments at all. Think carefully about whether what the person has said follows from the fact. Ask yourself whether they have established certain facts and whether their conclusions follow from their facts.

In practice

Remember that to counter any argument you can challenge the facts, challenge the conclusions, or find points that outweigh the conclusions. You've now learned the pitfalls and tricks of the trade, so work through each in a practice scenario so that you learn to recognize it when it is used against you.

Chapter
7

Golden Rule 7

Develop the skills for arguing in public

Many arguments take place in the course of a conversation, but sometimes there is a more formal structure to them. For example, you are arguing in favour of a proposal at a meeting. Or you are addressing a group of people. If that is the situation, here are some top tips.

How to speak well in public

- *Be prepared*. We have covered this in Golden Rule 1. In conversation you can get away with being a bit muddled, but if you're making a presentation have all the facts at your fingertips.
- *Practise*. Unless you are very experienced at public speaking, practise what you're going to say. It is amazing how an argument or joke works well on paper, but falls flat when spoken.
- *Talk slowly*. Probably the most common mistake in public presentations is that people talk too quickly. You might feel you're talking too slowly, but it's very unlikely that you are. Can you remember an occasion when you thought someone was speaking in public too slowly? I doubt it. But I bet you can remember an occasion when someone was talking too fast.
- *Don't read*. We've all been to presentations where a person simply reads a speech. It never works. It sounds stilted and awkward. Prepare bullet point notes. Have a list of bullet points to remind you of the structure of the talk. That said, it's always a good idea to have some notes to hand so that if something goes wrong you have something to help.

Avoid a speech with a beginning, a muddle and an end.

- *Smile at people*. Especially if you don't know your audience, try to arrive early to get to know some. It's really encouraging to see a familiar face in the crowd. Try to look at different sections of the audience as you're talking. Don't just look down or appear to be talking to one person. If you can move around a little, do.

- *Be brief and clear.* Remember Golden Rule 3. I've never been to a talk that was too short or too clear. I've sat through plenty that were too long and unclear. Remember the main thing is not to be funny or articulate or brilliant, but to convey the points you want to get across in a clear way. Make that your primary goal.

"Before I speak I have something important to say." Groucho Marx

- *Moderate the tone of your voice.* Use higher and lower pitches. Vary the pace. Use pauses. Speaking in a monotone can be disastrous.

Getting it wrong

A judge once said: 'Beyond doubt the dullest witness I've ever had in court ... he speaks in a monotonic voice ... and uses language so drab and convoluted that even the court reporter cannot stay conscious ... I've had it.'

- *Let the audience get used to you.* If you're going to be speaking for a while it's good to give the audience a little time to get used to you. Start with a story about your journey to the venue or something in the media, just so that people settle into your style. Don't feel it has to be a joke, a light-hearted anecdote is fine.
- *Quotes.* If you want to use quotes in a speech keep them short. Any quote of more than thirty words is likely to be a turn-off.
- *Use handouts.* These are especially useful if there is detailed information that the listeners need but that would be boring to go through orally, as in supporting statistics. They can also be useful to highlight your key points. It might be best

to give these out at the end of your talk so they are an aide-mémoire, rather than a distraction.

- *Using PowerPoint.* If you use PowerPoint keep the presentations clear and focused. Beware of the danger that people are so dazzled by your technology that they don't listen to you. I find that the use of a prop works far more effectively than a computer image. I use a water pistol in my criminal law lectures. Drenching the student gets across the finer points of different criminal offences better than forty PowerPoint slides.
- *Warning signs.* Watch out for warning signs that the talk is not going well. Are people fidgeting? Doodling excessively? Is there a 'buzz' in the audience? Are people checking their text messages? If so, don't panic, but do something! Do something unexpected. Have a story up your sleeve to tell, if necessary. Stop talking for a bit, that normally grabs people's attention! Stop and ask if there are any questions at this stage.
- *End with a clear, memorable summary of your argument.*
- *Questions.* If appropriate, ask for questions at the end. If you are asked a difficult question you can always reply: 'That's an excellent question. I can't give a detailed answer now, but I think we need to talk about that over coffee afterwards.'

Presenting in meetings

Here are a few further tips if you're going to make a short presentation at a meeting:

- If you know others who will be attending, talk to them beforehand. Try to get like-minded people on board. Check they are happy with the main points you will make.
- Encourage your 'supporters' to speak supportively as soon as possible after your presentation. Most people don't like to speak against a proposal if it's clear it has support, so this will pre-empt your detractors.
- As it's a short presentation, use a very clear structure. Say up front what you're arguing for, give three substantiated reasons (premises), say once again what it is you're arguing for (conclusion) and sit down.

Summary

Make it your priority to be clear and to be brief. Unless you are training to be a preacher or stand-up comedian you don't need to make people laugh or cry. You want to get your points across in a clear and convincing way. If you have a chance for a laugh on the way, all well and good.

In practice

The only way to develop skills in public speaking is to practise. You will make mistakes, but those will be great opportunities. Nearly all speakers will feel at the end of their presentations that parts went really well, and parts did not. So don't be put off if it's not perfect. Ask friends afterwards if they have any advice. Then practise some more.

Chapter
8

Golden Rule 8

Be able to argue in writing

Most arguments nowadays are done in conversation and discussion. However, with email and blogs, and in business and education, the formal written argument still has a role to play. Here are some key rules:

- *Write clearly*. Remember it's more important to be clear than to sound clever. You don't need to use long complicated words just because you're writing.

Don't be a sesquipedalian!

Yes, you guessed right. A sesquipedalian is a person who enjoys long words.

- *Use correct spelling and grammar*. But it's no longer a cardinal sin to start a sentence with 'but' or to use a proposition to end a sentence with. You're not writing a letter to your English teacher. Put clarity above grammar if necessary. Churchill famously replied to a tortuous, but grammatically correct letter, thus:

'That is the kind of language up with which I will not put.'

- *Think carefully about your opening words*. Readers will often decide whether to read on carefully or just skim a report based on the opening lines. You want something that will grab the attention and persuade the reader it's important to read what you have written. I once opened a book review with:

Are you a hirsute medical lawyer, with low blood pressure and writer's block? Then this is the book for you. You won't have to read too many pages before your blood will be boiling, you'll be pulling out your hair, and grabbing your keyboard to type a furious riposte.

I hope that grabbed some people's attention and made them want to read more.

- *Keep it brief.* Your reader is more likely to read a one-page summary than a fifty-page document. Remember the Ten Commandments are only 156 words. A lot can be conveyed briefly.
- *Use bullet points and paragraphs to separate out your points.*
- *Use the active tense.* 'I would have thought it best to proceed carefully,' would be better expressed, 'I think we should proceed carefully' or even better, 'We should proceed carefully'.
- *Read all the way through the argument after writing it.* Imagine you're the other person reading it. I remember meeting with a student of mine who was applying for a course at another university. He had sent me a copy of his application letter to look through. In the letter he explained that he was applying for the course as a safety net in case he did not get a job he was applying for. I asked him to imagine he was the professor running the course. How would he feel about receiving such a letter? I had to commend the student's honesty, but he wasn't thinking about how it would be read by the person he was writing to. Don't make that mistake.

The dangers of email

Email is an excellent way of communicating with people. It's quick and convenient. It can, however, be dangerous too. Please be careful when using email or blogs to argue.

How not to do it

The media love to report stories of 'email gone wild'. One notorious case involved two secretaries in a law firm. KN sent MB an email alleging that MB had stolen KB's sandwich from the fridge ('comprising of ham, some cheese slices and two slices of bread'). She demanded compensation from MB. It was all KN had for her lunch. MB emailed back that KN must have left her sandwich somewhere else. KN emailed back that MB was a dumb blonde. MB replied that KN could not keep a boyfriend. The exchange degenerated. Within hours it was being passed around the office and soon the partners of the firm. The secretaries both lost their jobs.

Nuance

One of the difficulties of email is that by its nature it is unspoken. This means that nuances that can be communicated orally are lost when written down. Imagine a colleague makes a suggestion and you reply:

‘What an interesting idea. Let's think about that in the New Year.’

You may have intended a really positive, encouraging reply. But the colleague could read that as if said with a sarcastic tone of voice, where what's really meant is that the idea is barmy and we shouldn't think about it again. When talking face to face you can normally work out if someone is being sarcastic or not by their tone of voice and their body language. All that nuance is lost by email.

The accent placed on a word can change everything. Contrast:

‘You want to go with that proposal?’

with

‘You want to go with *that* proposal?’

Emphasising *that* indicates that the speaker is astonished that the other person is enthusiastic about the proposal, but without the emphasis it appears as a straightforward question. Similarly, a remark intended as a joke can be read completely differently by someone else.

So when arguing by email, read carefully what you have written and imagine trying to read it in as negative a way as possible. Reword if necessary to ensure the message is positive. If you are at all in doubt, why not include a message at the end like: ‘I realize you might be reading this message and thinking I'm really annoyed with you. I'm not at all. We just have to get these things sorted out.’

Speed

I'm sure we've all done it. Sent an angry email and then moments later regretted what we've done. Or looked back at an email we sent the previous day and been horrified at how rude we were! Here are some top hints for avoiding that:

- If you're writing an angry email send it to yourself first. Read it again with fresh eyes. How would you feel if you received an email like that?
- If you're not sure whether an email is too aggressive, it almost certainly is! As a general rule, you will be coming across much more strongly than you assume.
- Remember there's a real person at the other end of your email – would you be happy to say this to them face to face?
- Why not send a draft of the email to a friend for any comments?
- Sleep on it!

Blogs

Blog forums have become a popular arena for arguing. Rightly so. They enable people who are interested in particular issues to get together and exchange views. When they work well they can prove a useful source of information and a ready way of finding out what others think. You can even see them as a way of enabling a large number of people to learn about your views.

But be careful! Remember that unlike conversations what you type is there for all to see, maybe for ever. The inaccurate statistic, the cruel response, the foolish point will be there for all to see and refer back to again and again! Most blogs enable you to post anonymously; this may be sensible and mean you can be protected if you say something you come to regret.

Blogs do seem to bring out the aggressive side in people. It's better to keep responses addressed to the arguments raised, rather than personal comments. Avoid swearing or clearly offensive remarks. They will not get you anywhere. I think some people who post on blogs forget they are communicating with other people. Most people are very sensitive and even a mere criticism can be blown out of proportion. So tone down your remarks. Just because someone has been rude to you doesn't mean you need to be rude back.

It's worth preparing what you're going to say in a separate document and checking it carefully before copying it over to a forum/blog. That way you can check that you have not made any obvious typos or said anything you will come to regret.

Summary

Learn to write in a clear and direct way. Don't try to sound clever or make things unnecessarily complicated. Use short, sharp sentences. Keep your writing brief and to the point.

In practice

Having written a letter or document, see if you can write it again in half the number of words. If you receive a letter you admire for its clarity, study it and see if you can learn from it. What was it that made it an effective piece of writing?

Chapter
9

Golden Rule 9

Be great at resolving deadlock

For many arguments I would recommend 'don't force a deal'. Often there is no need to get the person to agree with you immediately. Even though it might boost your ego to hear them say 'Ah, now I see how right you were and how wrong I was', there's no point in forcing them to that stage. Far better for them to think about it more and go through the arguments again. If they feel they've reached their own decision (rather than been browbeaten by you) they're more likely to stick with their new-found belief. Of course, they may decide on the spot, in the light of your arguments, to agree with you, but normally you don't need to force the issue. Give them time to think and be open for them to come back and discuss it further.

However, there will be cases, especially in the business world, where you'll want to push to a 'deal'. To explore this issue fully you might want to look at a business book on sales (try L. Thompson, *Mind and Heart of the Negotiator*, 2008, Prentice Hall).

Inertia

Most experts agree that the major problem in closing a deal is inertia. You can easily persuade someone that they would be better off with a new car or washing machine, or whatever it is. But it's then difficult to get to the stage of actually doing something about it. That's why magazines love people who take out standing orders or direct debits. Then customers need to take steps to cancel a subscription, rather than the journal having to persuade the customer to renew the subscription each year.

If you're in an argument and you need to close it, here are a few top tips:

- Create the impression that your offer is only open for a short time. Not for nothing do estate agents put up 'Sold' signs on properties in their windows. They want to create the feeling that if there is a house you like you must move fast or someone else will get it. If you're in a dispute with a builder try:

'Look, I want this business sorted out today. I'll pay you £150 if we call it quits now, but if you're not happy with that you'll have to take me to court.'

- Create the impression that everyone is, or is going to be, buying the product. Fear of being 'behind the times' or 'outdated' is an emotion many sales teams play on.
- Play on a person's self-image. Persuade the customer that being the kind of person they are, they should buy this product. I was stopped on the street recently by someone who said:

'You're the kind of person who cares about others, so please donate to help starving children.'

The impression was created that if I didn't give, I was showing I was the kind of person who didn't care about others. This ploy can be used with institutions or groups too:

'Do we want to be the kind of synagogue where this happens?'

'Is this the kind of community we want to live in?'

- Embarrassment can sometimes be an effective tool. Consider this discussion between exes:

Sue:	'Could you possibly have the children for a fortnight in March?'
Tom:	'I'm sorry, Sue, I'm afraid I can't.'
Sue:	'Could you manage them just for the first weekend in March?'
Tom:	'Well, I suppose I could do that.'

After rejecting Sue's first request, Tom will feel embarrassed by rejecting the second. If Sue had started out just asking about the weekend, he would have found it easier to say no. This technique can be used in quite a few arguments.

- Flattery, they say, can get you anywhere. That's probably an exaggeration, but it can certainly help:

> Brian: 'You did such a fantastic job on the stall last year, could you run it again?'

Moderation

The temptation in an argument is to settle on a compromise agreement. If the builder is offering to do the job for £200 and you're willing to pay £100, an agreement for a £150 fee seems inevitable. But don't be tempted into assuming that's what you should do. If you're convinced £100 is the correct fee, stick to your guns. If you believe the offer you're making is reasonable, don't be affected by the fact that someone else has countered with a more extreme offer. Indeed, they may well have made the extreme offer deliberately in the hope that you would increase your offer and so they would at least get a bit more.

Tip: Don't succumb to the argument that 'meeting halfway' is always the reasonable thing to do.

It's perhaps a particularly British characteristic to be drawn to the moderate middle road. But while that can be sensible on some occasions, be particularly wary of the temptation. Decide what you think is a fair figure. Consider this argument, which seeks to cut off complaints:

'I was going to ask for a 10 per cent pay rise, but I realize with the latest profit warnings that that's not reasonable, and so I'm only asking for a 5 per cent pay rise.'

This argument is clever because it makes it harder to suggest the person makes yet another sacrifice in their demand. Indeed, in face of the argument that the company is going through hard times, the person can present themselves as having already taken that into account.

What are your alternatives?

In any argument, don't forget what your alternatives are. And think about what the alternatives are likely to be for the person you're talking to.

Tip: Ask yourself where you will be if you cannot get an agreement.

If you cannot beat down the salesperson to a lower price, how will things look without the new car? If in fact the old car works fine then it's important to remember that in your argument. If they will not agree with your reasonable offer, then you have a good alternative: you can use your old car for a while longer. If your boss says no to the pay rise do you have other jobs you could go to? If so, you can push hard. If you don't have other job alternatives, you are going to have to ensure that the argument at least leaves you still with a job!

To give another example. Consider you are in an argument over buying a house and can't agree a price with the seller. What are the alternatives for you? How important is it you buy this house now? How important is it to the seller that they sell to you? If there are plenty of other people around who are willing to buy the house at a price higher than your offer, you have a weak case and there is little point trying to insist on your price. If there are no other buyers around and it is not urgent you move now, you can stick to your guns.

If you have a good alternative, let the other person know.

'That's fine if you're not willing to accept my offer for this car. I've seen a car in another garage that I'm interested in and I will see if I can do business there.'

What do you really want?

You might think you know what you want from an argument, but think carefully. What are your real long-term goals? Don't get your bargaining position muddled with what your basic interests are. You might think that selling the house for £400,000 is key to you. But how have you arrived at that figure? What are the long-term goals that led you to choose the figure? By considering these goals and focusing on them, other options might arise. If a person is seeking a pay rise, what is it they're really after? Is it increased status? Is it actually the money, or is it a competitive salary in comparison with others? There may be ways of meeting these demands apart from giving a straightforward rise. Creating the opportunity to work overtime, discussing freelance opportunities, increasing the kind of work done or changing the job title are all alternative solutions depending on the underlying cause of the request.

Deadlock

What if, despite all your arguments and discussion, there is deadlock? The temptation may be to walk away and leave the situation unresolved, but there are still alternatives left if initial arguments have failed:

1. Call in a third party. It's common in the business world to seek to rely on an arbitrator to resolve a dispute. Even if it's a personal matter, you could seek a mutual friend or trusted person to intervene. Of course ultimately the courts play that role, although if you can find some informal, cheaper alternative that may work well.
2. Secret bids. If the dispute is over payment, there are a number of devices that can be used. One that's popular is to suggest that both parties secretly put down their best offers. If they are within 15 per cent of each other then the average is taken, but if they are more than 15 per cent apart, then a third party will determine which is the most reasonable offer. Another possibility is to invite bids and use the figure closest to an agreed expert's assessment.

3. Flip a coin. It's old fashioned and simple but sometimes it works.
4. Take turns. In a famous court case disputing ownership of a large number of 'Cabbage Patch' dolls, the judge ordered the dolls be brought into court. The wife could choose one, then the husband one and so on until the whole pile was divided.
5. Famously King Solomon was asked to rule on which of two women was the mother of the baby. He ruled that the baby should be cut in two. One woman loudly protested and said in that case the baby should go to the other woman. King Solomon ruled that the protesting woman must be the mother and should have the child.

Summary

If you don't need to force a deal, don't. Give the other person time and space to think about what you have said. If you need to close with an agreement think carefully about what you really want at the end of the deal. If there appears to be a deadlock think laterally about whether there may be other ways of getting what you really want. If all else fails consider using one of the tie-breakers mentioned above.

In practice

If you have ever been forced into an agreement you didn't want, how did the other person do that? What could you have done otherwise? Always think of the big picture. Where will you be with or without this agreement in a year's time? Is this agreement part of a larger deal? If so, there may be no point damaging business relationships because of difficulties with one agreement.

Chapter

Golden Rule 10

Maintain relationships

You must see any argument in the context of the wider relationship between you and the other person. Before embarking on an argument you must consider what has been the relationship between you in the past and what will be the relationship between you in the future. What impact will the argument and the possible outcomes have on your relationship? There's much to think about here.

What is the argument really about?

It's important to remember that arguments are often about underlying issues, rather than the actual issue being discussed. There's a need to figure out whether the issue at hand really is what is being argued about, or whether the real issue is something else. Many people find that the argument they've had was in fact reflecting an underlying tension or difficulty. That argument about putting the socks in the laundry basket may in fact reveal a deep-seated concern about the relationship. In the business context the other firm may be appearing to be really tough, but is that because they felt hard done by with the last deal you negotiated? Or is their firm going through tough times, and if so what does that mean about how you should approach the deal?

What do you want?

Perhaps the first point to make is that very rarely does a person win an argument outright. It's unlikely after a discussion that your opponent is going to say: 'Do you know, I've been mistaken all this time and now I see you were right all along.' As a result, most arguments end when some sort of compromise is reached between the parties.

Do you want to remain friends?

In a witty article on how to win an argument, an American journalist wrote:

> "Suppose you're at a party and some hotshot intellectual is expounding on the economy of Peru, a subject you know nothing about. If you're drinking some health-fanatic drink like grapefruit juice, you'll hang back, afraid to display your ignorance, while the hotshot enthrals your date. But if you drink several large martinis, you'll discover you have STRONG VIEWS about the Peruvian economy. You'll be a WEALTH of information. You'll argue forcefully, offering searing insights and possibly upsetting furniture. People will be impressed. Some may leave the room." Dave Barry

It's so easy to win an argument and lose lots of friends. Take great care in how you argue.

Apologizing

There are times in an argument where an apology is needed. Perhaps it's clear you have behaved wrongly and there's nothing for it but to admit that and apologize. Refusing to apologize will make you appear big headed. If you need to apologize make it a proper one. That should involve the following elements:

- Clarity. 'I'm sorry that you feel I have treated you badly' is not an apology for behaving badly. Politicians are famous for creating apologies that are in fact, on close reading, not apologies. A proper apology must be a clear acceptance of the wrong done.
- A statement of what will be done to correct the wrong or an explanation of why it cannot be rectified.

These essentials should all be included where an apology is appropriate. But there are times when, although it should not be required, simply to move on some kind of apology would be useful. Something along the lines 'I can see that what I've done/said has really upset you and I would never want to hurt you and I'm sorry about that' might be appropriate. That acknowledges that hurt has been caused but avoids entering into a debate about who is right or wrong.

The importance of the relationship

In many, many situations the relationship is far more important than the argument. In the business context you might squeeze the very last penny out of your client by your fearsome arguing, but you may lose the client. Entering a deal that is fair and reasonable to both sides will produce a far more effective long-term business relationship.

As a consumer I remember countless garages where every time I went I felt suspicious that I was being ripped off, and I never trusted them. I now use a garage where, after several dealings, I have come to trust the mechanics. There have been several times when they did a minor task for free or their charges seemed moderate. They have a customer for life with me.

I'm sure we have all received good treatment from a firm and then recommended that firm to friends and thereby generated work for them. But that involves the consumer feeling happy with the agreement. Feeling they were forced into signing a contract or were browbeaten into an agreement is not going to do any good for customer relations and, in the long term, business.

I remember a situation where an employee was seeking a pay rise and pushed very hard for a salary increase a bit higher than I thought appropriate. Having given in, the next year when the issue of pay rises came around it was decided not to give him as big a rise as other people as he had done 'so well last time'. I later worked out that he would have been significantly better off if he had not pushed so hard the first time.

Losing an argument

You can't win every time! But first a word of warning. Earlier I urged you to be careful when arguing. Many people arguing with us are going to get something out of the argument if you lose. As we've seen, there are many devices that can be used to deceive you or wear you down. Here are just some:

- You may have been misled as to an important fact. Don't believe a statistic just because you've been told it.
- The argument that convinced you may have had a logical flaw.
- There may be arguments against the view presented to you that you have not been able to think of.
- You may have been so overtaken with the emotional appeal of the argument that you have failed to consider its merits.
- You may just be tired and not want to continue.

So don't admit defeat too easily, especially if that's going to affect you financially or cause you to lose your job. Unless there's some kind of emergency, no one should object to you saying:

'You've given me lots to think about and you make a very strong case. I need to go away and mull over the issues we've discussed.'

Indeed, if someone seems unhappy with that, you should wonder if they have something to hide. Are they worried you will find out a fact that will prove them wrong?

But despite all of this, it may be that you simply need to accept defeat. Many people try to save face in this situation.

'I'm sorry. I think I got confused about what we were arguing about. I thought we were arguing about X, but you thought we were arguing about Y.'

Others seek simply to end the argument without grace:

'I refuse to have a battle of wits with someone who is unarmed. Goodbye.'

That kind of remark might sound clever at the time, but it's hardly going to bring any long-term benefit. It is noticeable that Al Gore, in conceding defeat to George W. Bush, showed considerable graciousness and his reputation was enhanced because of that.

Winning an argument

Well done if you have won your argument! But be gracious in victory. We'll look at this more in Part 2 of the book. But if in winning you lord it over your opponent, you may win the argument and lose a friend.

Summary

Remember that keeping a good relationship with the person you're dealing with is more important than winning the argument. Perhaps you were not able to convince them this time, but there will be other occasions. Maybe you were able to convince them, but there may be other issues to discuss. Arguing can lead to a breakdown in relationships. Don't let that happen to you. Argue carefully and you will be strengthening, not weakening, your relationships.

In practice

Remember that relationships matter more than arguments. Whether you have won or lost the argument, you nearly always will want there to be a good ongoing relationship. If you're the winner then don't lord it over the other person: be gracious. If you have lost don't be a bad loser. At the end, whatever the result reaffirm the relationship. Spend some time together and just have fun. Go for a coffee and have a laugh.

Part
2

Situations where arguments commonly arise

Having set out my ten golden rules for arguing well I'm now going to look at some specific situations and how to apply them. We shall see that they can help whether you are arguing for a pay rise, arguing with your lover or arguing with the doctor. Perhaps not all of these situations will be relevant for you. But you are likely to came across most of them.

Chapter 11

How to argue with those you love

Arguments with partners and the extended family can be painful and complex. They can also go on for years! There's probably no area in life where it's more important to argue well. Perhaps it's of some comfort that you'll have plenty of practice at this! I don't know a couple who find they never seem to argue enough.

Getting it wrong

Shamrita:	'You've left your socks lying around again.'
Suni:	'Well, I'm not stopping you picking them up. And while ...'
Shamrita:	'What do you expect me to be, your servant?'
Suni:	'Well, you seem to like treating me like a child.'
Shamrita:	'If you grew up a bit I wouldn't need to!'
Suni:	'Typical, I do all the work and bring in all the money. You just lounge around the house all day with the only thing stressing you being my socks. You need to get a life!'
Shamrita:	'Well, maybe I should. I'm trapped living with you. I need a life. Leaving you will be a good first step.'

This example shows how easy it is for arguments over the most trivial things to get out of hand and escalate out of all proportion to the initial issue.

In fact, arguments can play a beneficial role in a relationship. They can help each party realize what the other really cares about. An argument enables there to be an outlet for feelings of antagonism that might otherwise fester. All relationships need boundaries and limits. A relationship in which one party got their way on every issue would be a bad relationship. If a wife constantly submitted to her husband that would be disastrous. An English judge once said that 'on marriage a husband and wife became one and the husband is that one'. That's now an outdated and unacceptable model for a relationship. Relationships are about give and take; mutuality, if you will. Arguments are where you need to determine how your competing interests can be balanced.

"Most couples have not had hundreds of arguments; they've had the same argument hundreds of times." Gay Hendricks

How to argue with your partner

Here are some top tips for arguing with partners:

- *Remember Golden Rule 2*: Choose your time and place of arguing. You will know the foibles of your partner. My wife knows that if I'm hungry then it's not a good time to raise a sensitive issue! If there's an important issue that needs discussing try to choose a time when you are both relaxed and have the time to discuss it. I know ... that will be never! But at least do your best.
- *Remember Golden Rule 3*: It's not what you say but how you say it. Don't lose your temper. I touched on lots of tips on how to keep cool when that rule was discussed. If you feel you're getting angry then put space between you and your partner and calm down. Not only is losing your temper not good for your relationship, it's bad for your health.
- *Never ever be violent*: Never hit, throw things or physically threaten your partner. If you fear you might become or you have been violent, seek professional help quickly. If your partner has been violent to you think very carefully about whether it might be best to leave the relationship. All the evidence shows that those who are violent to their partners are so repeatedly. Typically, a violent partner is deeply apologetic, only to return to more serious violence in the future.

- *Remember Golden Rule 4*: Listen, listen, listen. It's only respectful and proper to listen carefully to your partner. Don't interrupt. Don't finish their sentences. Where your partner has made a fair point then accept and acknowledge it. Demonstrate that you have heard what they have said and you accept it. So often in arguments between partners each is so keen to list the faults of the other, they don't listen to what the other is saying.
- *Try to make statements in terms of what you feel*: 'I sometimes feel that you care more about your job than you do about me' is less confrontational than 'You care more about your job than me'. In putting the argument in terms of what you feel, you're not judging the other person and are just describing your emotions. It opens the door to a ready reconciliation: 'I'm so sorry you feel like that. Of course, I care far more about you than my job. I know I've been working late recently but ...'
- *Focus on the future, not the past*: Normally with relationships there's not much point focusing on the past. How should the issue that is causing friction be dealt with in the future? It may help not to criticise, but rather to request. 'In future, could you load the dishwasher after lunch' may be more effective than 'You never help with the washing up'. Focusing on the past produce an apology and shame, but it might also lead to personal insults, frustration and anger. Focusing on the future brings a solution to the issue without harming the relationship.
- *Remember to see the issue from your partner's side*: Acknowledge the benefits your partner brings to the relationship. 'I know you have the baby to look after all day and there's nothing more exhausting than that, but ...' Acknowledge the good. Make it clear that you do love them and respect them.
- *Take time*: If you're aware your discussion is leading to an impasse, suggest taking time to think about it. There might be more options that you can't see. And make it clear to your partner that you're not trying to avoid the issue. Promise to return to the discussion at a set point, perhaps 'tomorrow morning after I've had some sleep'. Sleeping on an issue can

also provide perspective. I'm sure we've all been in the position of waking up 'the morning after' and wondering why there had been an argument over an issue that seems so trivial the next day. At the time it might have seemed important where the toothpaste should live, but it all looks rather silly the next day. Beware, however, of putting things off forever. That might just disguise underlying issues.

- *Set a time limit*: If you're having a discussion it may be worth setting a time limit, agreeing to return to the issue later if you have not resolved it by then. Decide to do something fun after that.
- *Be alert to what the real issues are*: In relationships it's common for a trivial issue to lead to a major argument, but the trivial issue can reflect a significant point. The dispute over the toothpaste may reflect a broader issue that one person doesn't think the other respects them; or that one party feels the other is trying to control them. We can see this in the example of the argument between Suni and Shamrita. At first it appeared that there was an argument about socks, but there were clearly a lot of other issues going on here. Suni felt that Shamrita was always trying to control her. Shamrita seemed to find her life based at home unfulfilling. These are major issues. If they're not resolved their relationship could be doomed. First, they need to sort out the sock issue rapidly. The bigger issues will require a longer, more serious conversation, probably at another time when they have plenty of time to discuss them.

Suni: 'Well, I think it's good we have aired these issues. I think we need to take time to discuss them further. On the issue of socks, I'll try to remember to put them in the laundry bin. I'm often in a hurry to get to work and forget, but I will try and do better. But maybe we should spend some time together tomorrow discussing where our lives are at more generally.'

Reconciliation

- *Remember Golden Rule 10*: Maybe your partner is being un-reasonable or making a demand you think of as petty. Still, your relationship matters much more than a trivial issue. If your partner feels something is important, you should respect that, even if to you it's a trivial issue. In our situation, is Suni so desperate to be allowed to leave her socks on the floor that she's willing to endanger her relationship with Shamrita? If taking that little bit of extra effort with the socks means the relationship continues, isn't that worth it? Don't get misled into thinking about 'my rights': your rights may be important in grand political debates, but with relationships it's about what makes you work as a couple.
- *Be forgiving*: When you're with someone all the time you'll see them at their weakest and most vulnerable. You'll see them when they're exhausted and frustrated. We all need times when we can let our guard down and not put up pretences. Partners will see each other when they do that. So you can't expect perfection from yourself or from your partner. You must be forgiving and understanding.
- *Be ready to apologize*: As we have just seen, you can't expect your partner to be perfect and you can't be expected to be perfect. Be very ready to apologize. It's amazing how a quick 'I'm really sorry. I shouldn't have said that' can change a situation that could have led to a major row into a pleasant evening. It costs nothing to apologize. A failure to apologize will leave your partner feeling that you haven't understood their feelings and you don't care. Note that an apology for hurting someone's feelings or saying something cruel doesn't mean you are losing the argument. The issue that caused you to misspeak can be returned to once the tension has abated.
- *Be positive*: If you have had an argument, try to make sure good comes from it. Otherwise you will have the same argument repeatedly. Having a positive outcome will nearly always involve both parties agreeing to change their behaviour. Suni needs to learn to put her socks away and perhaps

Shamrita needs to learn not to criticise every time Suni forgets. And if you have made an agreement, do your best to stick to it.

Getting it right

Shamrita: 'You've left your socks lying around again.'

Suni: 'Look, I'm really sorry about that. I was in a real hurry this morning. Sorry you've had to tell me about that. That must be really annoying.'

Shamrita: 'OK, then.'

Suni: 'It seems I'm often doing things that annoy you. I wonder if it would help if we spent some time tomorrow evening talking about this?'

Shamrita: 'That would be good. Do I tell you off that often?'

Suni: 'Well, it does seem like that sometimes. We can talk about it tomorrow. Let's go out for a nice meal and have some quality time.'

Summary

It's crucial that you get arguments in relationships right. Arguing well is a good part of a healthy relationship. Treat your partner with respect and listen carefully. Remember that matters that seem trivial to you may matter a lot to your partner. Talk through issues together and work out a solution that will work well for you as a couple.

In practice

Here are some useful phrases:

- 'I'm really sorry I've upset you. I love you very much and never want to hurt you. I think we need to take some time to talk this through. Let's go for a walk tomorrow by the river and talk about this.'

- 'I know you don't mean to upset me, but when you say things like that it makes me feel you don't really respect me.'
- 'Look, I think we have a problem here. I realize that your football is really important to you and it gives you great fun. But it means I'm left with the children most of Saturday and I end up feeling I get no time to myself. Can we talk about a way we can deal with this?'

Chapter
12

How to argue with your children

Why is it that children can be more exasperating than anyone else? Most parents will have despaired of their children at some point:

'They just don't listen. I can't get them to do anything. It's non-stop arguing.'

Perhaps we shouldn't be surprised. Parents inevitably have to treat children in a way they wouldn't dare treat anyone else. Have you ever tried telling an adult they really ought to go to bed, or that their clothes don't match? We don't like being told what to do and it's no surprise that children don't either. Remember, children have their rights too. The ten golden rules apply just as well to children as they do to adults.

Getting it wrong

Dad: 'Steve, you're not going out until you've done your homework.'

Steve: 'Look, Dad, I'm 15, not 7. I'll do it later.'

Dad: 'Listen. I'm your Dad and you do what I say.'

Steve: 'OK, I'll do my homework tomorrow morning. I have to go now or I'm going to miss the party.'

Dad: 'You go to the party and there's no pocket money for the rest of the month.'

Steve: 'Yeah. There's no way you'll do that. OK, I'm off.'

Dad (grabbing Steve's arm and shouting): 'You do what I say. You're not going anywhere.'

Steve (pushing Dad away): 'Let go, Dad.'

(Steve pushes Dad away and leaves)

This is typical of all too many interactions between teenagers and parents. We'll come back at the end of the chapter to see how this argument might have gone so much better.

Tactics

Here are some of the tactics that parents commonly use in arguments with children:

1. *Threats*: 'Do your homework or there will be no pocket money this week.'
2. *Rewards/bribes*: 'Do your homework and you'll get an extra £2 pocket money this week.'
3. *Logic*: 'Do your homework and you'll get better marks in your exams.'
4. *Power*: 'Do your homework because I say so.'
5. *Guilt*: 'We have done so much for you, surely the least you can do is do your homework.'

There's nothing inherently wrong with any of these tactics. But they need to be treated with care. We'll look at them one by one.

Threats

Threats are a major weapon in the parental arsenal! Parents can easily control children's access to things they want. In the case of smaller children, parents can even physically impose their will (e.g. by carrying them to their rooms). However, threats need to be used carefully and can be misused:

- Don't make threats you don't intend to carry out. Your child will soon learn that you don't carry out your threats. Indeed, older children will readily see you don't intend to carry them out.
- Start with lesser threats before increasing the level of threat. Start with, 'I'll have to consider whether to cut your pocket money ...'

- Be proportionate. Don't make threats that are out of all proportion to the wrong being committed. Most children have a strong sense of what is fair.

In many cases it's better to put the threat in terms of a choice for the child.

Useful example

'You can either choose to tidy your room and get your pocket money, or choose not to tidy your room and get no pocket money.'

One of the benefits of expressing threats in terms of choices is that it teaches children that there are consequences for their actions. This is a lesson they must learn in life. It also empowers them to choose between the consequences. Of course, such a tactic is only sensible if you're prepared for them not to tidy their room and to withdraw their pocket money as a result.

Rewards/bribes

This is perhaps the favourite among parents. It always feels better to give a reward rather than a punishment. Again, there are dangers:

- Try not to get into the habit of using bribes all the time. There should be some things that children do as a matter of course. Keep bribes for dealing with unusual situations (when the train is cancelled, or the child is screaming in the restaurant).
- If you are offering a bribe or a benefit make sure it's close in time. A promise of a benefit next week is unlikely to be as effective as a benefit when the task is performed.
- Make sure your bribe is proportionate. Don't use big bribes to get a child to do something straightforward.

The main disadvantage of bribes is that a child can easily learn that the best way to get nice things is to be badly behaved. That way you are given a bribe to be good. It's very important therefore to use rewards more than bribes. If a child has behaved well and tidied their room as requested, give them a reward afterwards. As many experts say, if you pay attention to bad behaviour and ignore the good, you're in trouble! Yet if the child is behaving well the temptation as parents is to ignore them and get on with your own things.

The other benefit from using rewards is that it reinforces the lesson that actions have consequences. As we have said already, a crucial lesson for children to learn is that what feels good now might not be the best thing to do. Doing something less pleasant now may lead to benefits later. And doing what is pleasant now (eating a whole tub of ice cream, say) may be regretted later. Mind you, that is a lesson some of us are still learning!

Logic

Of course, not every child or every situation can be dealt with by logic. Younger children or children who are very upset cannot appreciate a carefully honed argument. And there may just not be time. But where possible use logic and sound arguments with your child, if for no other reason than it will be cheaper and less exhausting than using bribes or threats. More importantly, using logic will teach your child how to make decisions and how to think for themselves.

Use arguments that will fit in with the way the child sees the world. Telling a 7-year-old to do her homework so that she will get a place at a good university is unlikely to work. The same argument for a 17-year-old who is desperate to get to a particular university might well be successful. Don't assume that the arguments that are persuasive to you will be persuasive to your child. You might think 'You will get really cold wearing that outfit' is an overwhelming argument, but for the child it might not be. Similarly, 'That's what all my friends do' sounds like a terrible argument to an adult, but is very powerful for children.

This is where listening is important.

Useful example

'Why do you think that's a good idea?'

You have to find out what is motivating the child to make their decision. Can you work within that way of thinking to get what you want? So the child who wants to wear the skimpy outfit because they want to look trendy might agree to wear a coat until they get close to the party.

Power

This is rarely an effective way to win an argument. The older the child the less likely it is to work. In any event, it's not educational. The only lesson it teaches is that if you are stronger than someone else you can and should impose your will on them. Preaching at children, even if you get them to do what you want, might in the long run alienate the child. It's so much more effective to persuade the child to think for him or herself and thereby reach a sensible conclusion.

All of that said, there are times when there is no choice but to exercise power. If the child has to get to a hospital appointment and is refusing to go out of the door, there may be no time to do anything other than pick them up and put them in the car. But when that's done, discuss the issue with the child afterwards. Talk about why you had to pick them up.

Guilt

Guilt is an argument commonly used by parents. All parents make great sacrifices for their children and some children seem particularly ungrateful. What parent has not thought that their child has no idea how lucky they are?

But using guilt is not productive. Remember Golden Rule 10: the long-term relationship is key. A relationship that is built on guilt and a sense of obligation is not likely to be a beneficial one in the long term. There is a place for reminding children how much more advantaged they are than others. But reminding them about all that you've done for them is not normally a good idea. Anyway, we all know what they will grow up to say: 'I didn't choose to be born!' Reminding a child of all you have done for them may just create resentment and ignores the real issue at hand. If a child simply does what you say due to feelings of guilt that will not be the basis of a beneficial relationship.

For example, imagine your child is sulking in a toyshop when you're out buying a present for his friend's birthday party because you won't buy him a present as well. The temptation is to say: 'You don't know how much I've done for you, and yet you always want more! You don't know how lucky you are!' This might very well be true, but when a child is sulking in the middle of a toyshop he is unlikely to respond to this kind of argument. It would be better to say: 'Thank you for showing me that toy you want. Next time we buy you a toy, that would be a great one to get. Today it's not your turn for a toy, but if you have good behaviour, we'll think about getting you another toy soon.'

General principles for children

There are some important general rules in relation to children:

- *Don't use corporal punishment*. Most of the experts in the field (including paediatricians, social workers and academics) believe that corporal punishment is ineffective and harmful.
- *Stay calm*. Yelling at your child might sometimes be effective in getting a short-term goal, but in the long term it teaches them that yelling is appropriate and they will use it when they lose control. Parents should, when possible, model good behaviour! Of course, all parents do shout at times, and you would hardly be human if you did not. But keep that to a minimum. If your child is being exasperating simply take time away. Walk away and take a breather. Have a drink of water. Ask your partner to deal with them. Indeed, it is surprising how often introducing a new person into the situation can lead to a rapid solution.

- *Praise your child.* Even when correcting them, emphasize the good things they do and encourage them to do good in the issues you're arguing about. Remember too to give rewards and encouragement for good behaviour. If good behaviour gets no response and bad behaviour leads to a telling off then bad behaviour may become the only way a child can get a response from you. You must learn to recognize what might be attention-seeking behaviour. Children can be provocative just because they are tired and need some attention. Stopping the argument and giving them a cuddle can be very effective. Of course, if this is the issue, then build quality time for them into your schedule so that they don't need to resort to bad behaviour to get your attention.
- *Treat your child with respect and as an intelligent person.* Give them reasons for acting in a sensible way. Listening to the reasons they have for not complying with what you say can be important. You might decide, when you listen properly to them, that they have a point after all. Remember that the things that are important to children are not necessarily the things that are important to adults. You shouldn't expect them to be little adults, but good children. All of this will give your child invaluable lessons for life as they learn to think through issues for themselves. By treating your child with respect and as an intelligent person, they are more likely to treat you (and others) in the same way.
- *Spend time with your children.* You can only get to know your children well if you spend time with them. Only then will you know what triggers disputes and what kinds of reasons they are likely to listen to when arguments arise. There is plenty of evidence that good relationships with parents can benefit children in terms of education, psychological well-being and happiness.
- *Consistency is key.* If you have rules, stick to them. If there are rewards or punishments that follow from behaviour, then respect them.
- *Be careful what phrases you use.* Adults are accustomed to unpleasant phrases being used and are normally able to ignore them or put them in context. Children find this much

harder. 'You're stupid' might easily be laughed off by an adult, but not by a child. So be particularly careful of personal attacks on a child. Concentrate on remarks about their behaviour rather than them. This is extremely important. Self-esteem issues that develop in children can grow into much bigger issues later. So address the issue/behaviour at the heart of the argument, but don't attack the person. Say 'Writing on the wall was *not* a good thing to do!', rather than 'You are a naughty and silly boy!'

Useful examples

'That behaviour is inappropriate for someone of your age' (rather than 'You're being babyish'). For a younger child, say 'That's how a two-year-old would act. You are now three.'

'I don't like it when you use language like this' (rather than 'I hate you').

'You're a very clever person, but you don't sound it when you talk like that' (rather than 'You're stupid').

- *Remember that children learn how to argue from you.* Speak harshly, fail to listen, be abusive, shout, and children will learn that is the way to argue.

Unruly children

Perhaps the most difficult situations arise where children get angry and you're trying to discuss things with them. First, remember that anger is a normal and natural emotion. The difficulty for children is often how to deal with anger. Don't make the mistake of thinking that a child is being bad because they feel anger. The problems may be with how that anger manifests itself.

Here are some top tips:

- Find out why the child is angry. What is it that riles the child? Is it something that can easily be dealt with? Some children become short-tempered when they are hungry or tired. Perhaps a quick snack is the answer. Making sure the child gets enough rest may be key. Is there something that you do that creates the anger? Remember, it's so easy to see the situation only with adult eyes. To you the fact that the child is upset because he has lost teddy, when he has dozens of stuffed toys, makes no sense, But that is not how children see the world.

- Tell the child you understand their anger. Do your best to empathize with the child.

> **Useful examples**
>
> 'I can see that you are really angry.'
>
> 'You know, I get angry sometimes too. I can see you're really upset.'
>
> 'When Sue did that to you it was really really annoying, wasn't it?'

- Acknowledge that the child is angry and, if you know why they feel that way, acknowledge the wrong that was done to them. Depending on the age of the child it may be appropriate to help them name the emotion they're feeling. Talking through anger and emotional responses with children can be immeasurably valuable. Coming alongside them and being a friend, rather than confronting them when they are in an angry state, can build your relationship. The issue you're concerned with can be dealt with later when the child is calm.
- Show the child a good way to express their anger. This is best done later, when the child has calmed down. Ask yourself what should he or she do when they feel anger? Maybe you should encourage the child to go outside and do something energetic ('When you feel annoyed why not go and ride your bike?' 'Why not hit a pillow when you get really angry?'). My wife tells our daughter that she is allowed to be as cross as she wants to in her room, and stamp around as loudly as she wishes, as long as it's in her room. You have to give your child

a way to work through their anger and frustration, and help them recognize anger as a normal feeling. Teach your child that it is what they *do* in their anger that can be harmful to them and others, and give them healthy ways of expressing it.

- Listen to what the child is saying and make sure your child realizes you are listening. Repeat back to them what they have said so that you have understood correctly. You need to teach them that listening to someone else is important. Why should they listen to you if you don't listen to them?
- It is tempting sometimes to see any anger issues your child might have as 'their' problem. But it's better to see it as your family's problem. Indeed, the school and friends may all have their part in working through it.
- When a child is feeling angry or emotional is not the time to have a productive argument. Nor is it the time to correct the child. You will need to address the issues at hand later on when they are calmer.

Useful example

'This morning, when Sylvie changed the TV programme you were watching, you got very angry with her and said some nasty things to her. You really need to go and apologize to her.'

- It may be that getting on the same eye level with the child will help. Or maybe coming side by side to them. Each child is different, and so you need to find the best way to communicate that you're with them and will help them. Use a calm voice and a pleasant expression. Depending on the child, holding them or touching them may be helpful. Other children will want space and will not want to be touched while they are angry.
- Keep things short. Children don't want (and it's rarely helpful) to talk at length. Deal with the issue quickly. There may be another time to talk in more depth.

- It may be that there are medical issues connected to your child's behaviour. A chat with your GP may reassure you or they may offer tests if you're concerned that there is more going on than normal child behaviour.

Teenagers

Many of the principles we have already discussed are applicable to teenagers. Here are some of the key points:

- Spend time with teenagers, but be aware of their desire for privacy and space. You may need simply to be there and available, even if not directly talking to them. When discussing, try to use open questions, 'How was your day?', rather than closed ones, 'Did you have a good day?', which can be answered 'yes' or 'no'. If your teenager is keen to talk use that as a time to chat, rather than a time to reprimand. Arguments can take the form of 'discussions' in this context.
- Listen to their arguments and respect them. Try to respond in arguments that they will appreciate. For example, the fact that you and your friends might think a particular outfit is unsuitable is not going to persuade a child who thinks it's trendy. Indeed, for many teenagers being able to choose what to wear is a crucial aspect of personal identity and independence. That must be respected and understood.
- Many teenagers suffer from self-esteem issues. Be sensitive to this, particularly in relation to matters of personal appearance. Don't criticize them harshly or make derogatory remarks. Even jokes can be taken wrongly. Build your child up. Respect them as young people.
- Some teenagers find dealing with and expressing emotions difficult. You need to understand and support your teenager during this time. Gently help them express those emotions in suitable ways. Telling them off or punishing them is unlikely to help.
- Remember that for many teenagers their status with friends is very important. Chiding them about their untidy room in front of their friends is unlikely to be popular. It's best to deal with any problems when you are alone with your child.

Getting it right

Dad: 'Where are you going to Steve?'

Steve: 'I'm off to a party. I've got a lift, but I have to go now.'

Dad: 'But have you done your homework yet? When is it due in?'

Steve: 'Tomorrow.'

Dad: 'So how are you going to get it done?'

Steve: 'I'll do it tomorrow morning.'

Dad: 'Are you sure you'll get up in time?'

Steve: 'I should do.'

Dad: 'Well, you're going to get in trouble with the school if you oversleep.'

Steve: 'That's true. Look I'll make sure I'm back by 11 and I'll set the alarm clock for 7.30.'

Dad: 'OK, but we'll use this as a test case. If it doesn't work out this time, do you agree that in future homework will have to be done before you go out? You can only go out now if you agree.'

Steve: 'Fair enough. Bye Dad.'

Summary

Children are great! Encourage your children to be good, rather than overreacting when they are bad. As far as possible try to reason with them and give them good reasons for acting in the way you want. Discuss with them why things go wrong. Help them to learn the consequences that flow from decisions. Always love them. Very much.

In practice

Talk to your child as much as possible. What makes them tick? What kind of person do they try to be? What do they enjoy? Try in arguments to build on these aspects. Try as much as possible to work through issues together with your child, rather than ordering them what to do.

Chapter
13

Arguments at work

Do you keep getting into arguments at work? Do you find it difficult to stand up for yourself? Do people keep telling you what to do and you either end up in an argument, or you roll over and feel you're a doormat? Are you a boss and do you often find yourself in arguments with employees? This section will give you some key strategies for arguments at work.

Getting it wrong

(Monica and Jessica are at a meeting at work with colleagues)

Monica: 'I would like to propose that we go ahead with this deal.'

Jessica: 'What? Like the proposal with that disastrous Birmingham deal a couple of years ago?'

Monica: 'Well, let's put that behind us. Just like we have your hopeless firm outing last year.'

Jessica: 'OK, where are your figures for this deal? Not that we will believe them!'

Monica: 'There seems no point in discussing this.'

Jessica: 'How right you are. I would keep quiet if I were you.'

Boss: 'Now, Monica and Jessica, I think you both need to calm down. This is not reflecting well on either of you.'

Avoid arguments where possible

Workplaces can be intense places to be. Stress levels can run high and it's easy to lose your temper or get into an argument you later regret. Follow Golden Rule 2: is this argument really important? Is this the time and place for the argument?

Perhaps there is someone in your firm you find you are constantly arguing with. Avoid them! Or try to make sure projects you're involved with don't include them. Or better still, try to meet with them so you can reconcile your relationship and be on a more equal footing.

Be an encourager and praiser. Being positive in the workplace will make it easier for you if there are things that need changing. People will also listen to your complaints if they know that generally you are on their side and supportive of them.

The time and place

We looked into this issue when we discussed Golden Rule 2. But the questions you should ask yourself are especially important here. In particular:

- Is this an argument best had at a meeting, or privately?
- If a private meeting is better, do you want someone with you or will it work best one to one?
- Is this matter best addressed on paper or face to face? If face to face, will it help to send an email first, setting out your concerns?
- Can you imagine a better time to have the argument? For example, Friday afternoon at 4 p.m. might be best avoided.

If you are a junior employee or new to the firm it might be best to check that this is a good time to raise your concern:

‘I do have a couple of questions about this proposal. Is this a good time to raise them?’

Be seen to put the business first

Many companies are full of people with considerable egos. Sometimes competitiveness between co-workers is encouraged. But you should at least be seen to put the business first. Put your arguments in terms of what will work best for the company, not what will work best for you. By putting your arguments in terms of what will help the company you can draw in support from other workers. You can also start to find some common ground. Hopefully everyone will agree with you that promoting the company will be best. Indeed, if you can present the person you are arguing with as putting their own interests before the company's you are well on your way to winning your argument.

Also, check why it is you are raising this issue and creating an argument. Is it just an attempt to look good? Or make someone look bad? Or is it a really important issue for the company? Be a bit suspicious about your motivations. There can be serious fall-out from arguments with bad motivations. Avoid them.

Pick your arguments carefully

In business life there is plenty to argue about. There are lots of things that you probably think could be done better. But if you gain a reputation for being the kind of person who argues about everything then your points will lose their strength. Leave that argument about the coffee-maker for others, and wait for the important ones. The person who speaks little is often listened to very carefully when they do have something to say. The person who is arguing all the time is easily greeted with 'here we go again'.

Encourage discussion

If you are in a management position there can be a tendency to discourage discussion or arguments. That's not always a good idea. The more people voice their ideas and concerns the better, at least on major issues. You want concerns and issues raised and dealt with now, rather than when the proposal is well developed. It's especially tempting when chairing meetings to push material through. But while speed is important, it's even more important to make the correct decisions. Resentfulness can easily build up among a workforce if it's felt inappropriate for them to say what they think. Having people on board who are secretly opposed to your plan is unlikely to be productive. Similarly, if a tradition of aggressiveness at meetings where there is disagreement develops, this will deter people voicing their views. Encourage a respectful listening to all views.

This is why terms such as 'brainstorming' can be useful. The understanding is that all kinds of ideas will be thrown around, without the impression that this is an argument that people should get worked up about.

Get others on your side

If you're planning a meeting or confrontation, get others on your side. Discuss the issue with colleagues in advance. You might find out who is likely to oppose you and why, but more importantly it will strengthen your position to know you have supporters with you in the meeting. If, after your presentation, there are immediately people queuing up to say they agree, this will strengthen your position. You may find for someone there is a particular issue that concerns them and you can win them round by making a fairly small concession. This is much easier to do in advance of the meeting than it is during a meeting.

Never, ever, lose your temper

Losing your temper at work is disastrous. It will appear unprofessional and uncontrolled. If you are prone to lose your temper use all your skills to avoid this. See Golden Rule 3. If you're heading for a stressful meeting and you can foresee yourself getting angry (perhaps there is a particular person who always gets your goat) then imagine yourself being provoked but keeping calm. Thinking through situations in advance and controlling your temper is extremely important. Your reasoned arguments will be heard if they are presented calmly and professionally. Losing your temper will most likely lose you the argument.

Get issues resolved

The temptation with arguments is to find a compromise and push ahead. Sometimes that can work. But be careful. A compromise may paper over fundamental differences in approach between two factions in the firm. They may both sign on the dotted line of the project, but have in mind fundamentally different visions. It's important to resolve these differences if at all possible. However, it may be that there is no resolution and one course has to be taken. Then do all you can to keep everyone on board. Make sure they at least feel they have been listened to. Acknowledge what they said and say how listening to all the arguments has helped produce the best decision. Where possible indicate their concerns, and show how you have taken them on board. Find positive things for them in the deal.

Be honest

If you are desperate to see your project promoted or your view taken on by the company there can be a temptation to massage the figures or disguise potential problems. Never lie. If you are found out it can be the end of your career. You may never be trusted again. It's not worth the risk.

Healing after an argument

If you have had a row at work it is best to try to resolve it quickly. If you have behaved improperly then apologize and seek to reassure the other person that it won't happen again. This is particularly important if you have been rude to your boss! Being honest and professional with your boss will stand you in the best place to keep your job.

If you have had a row at work and you are the boss, you still need to mend any relationships that might have broken down because of the row. If you don't, factions will unite against you and you'll find it more difficult to exercise authority as the boss. Restoring harmony in the office is important, so do your best to smooth over differences and move forward. If you have behaved badly, apologize for the behaviour. Apologies are not a sign of weakness. It takes a strong person to apologize. You will be more respected by your employees if you hold yourself accountable for your actions.

Getting it right

Monica: 'I would like to propose that we go ahead with this deal.'

Jessica: 'I think it's important we think about this very carefully. It might be really good for the firm, but we have made mistakes in the past and have to be really careful.'

Monica: 'Yes, I remember your hopeless firm outing last year.'

Jessica: 'Monica, I think we need to focus on the issue at hand. You have set out very helpfully the benefits of the deal. But we need to consider the dangers.'

Monica: 'OK, how should we do that?'

Jessica: 'Well, there are two "nightmare" scenarios that we need to think about ...'

Summary

Choose your arguments in the workplace carefully. Where possible make sure there are others on side with you. This is especially useful in meetings. Make your points clearly and respectfully, with the focus being on the well-being of the company.

In practice

Spend a meeting just watching what others say and do. Which interventions are useful and which are not? How do people get their proposals passed? What sorts of arguments sway the management in your company? If you're a manager, you might ask if you're getting a good range of opinions from your workforce.

Chapter
14

How to complain

It has happened to us all. The product looked excellent in the store but fell apart when we got it home. The electrician who seemed so reliable when we first met him has failed to complete the job even though he has been paid. The holiday we bought based on the brochure was not at all what we got when we arrived. In all these situations, if we are to get our money back an argument seems inevitable (although, if truth be told, I'm sure there are times when we have felt not up to it and have written down the loss to experience). But if we do choose to argue, it can be hard to get our point across even though we are in the right. This chapter will show how complaining about faulty goods or bad services need not be stressful.

Getting it wrong

Jonathan: 'Hello, is that Clogs? I bought a pair of shoes from you yesterday and they fell apart as soon as I got home. I'm outraged. This is appalling service. In fact, never in all my life have I seen such shoddy merchandise.'

Tina: 'Good morning sir, can I ask how the accident ...'

Jonathan: 'Are you saying that I broke the shoes? That's just typical. It's your shoes that are faulty, not me!'

Tina: 'I just need to know when you bought the shoes.'

Jonathan: 'Right, I'm fed up. If I have to fill in some lengthy questionnaire just to get back my money I'm not going to. I want to speak to the manager. Now!'

Tina: 'Well, I'm afraid the manager is busy just now, but ...'

Jonathan: 'Liar! You know he's not and I know he's not. Can't you treat me like an intelligent human being?'

Tina: 'Sir, I'm trying to help you...'

Jonathan: 'Yeah, right.'

(Jonathan slams down the phone)

Fairly obviously, Jonathan has got nowhere with this complaint. Applying our golden rules is important in this sort of situation.

Avoid the argument, where possible

Remember Golden Rule 2. With many consumer problems the best solution is avoidance. Buying products that are well-known brands from reputable shops should decrease the likelihood of goods not working. There are plenty of websites and publications that can give advice on the reliability of certain products.

With regard to employing workpeople to do tasks, nothing beats a personal recommendation. If a plumber has done a good, reliable job for a fair price for your friend then they'll probably do the same for you. Even then, it pays to play safe, especially with larger jobs. Here are some top tips:

- Always have a contract. Agree in advance precisely what job will be done and when payment will be made.
- Never pay the final instalment until the job is completed. If the builder refuses to agree to such an arrangement, be very suspicious. Keep enough money back so that you could pay someone to complete the job.
- Avoid paying large sums up front. It may, however, be reasonable for you to pay for some materials up front.
- In the case of larger projects, such as extensions, build in a sum to pay for 'tidying up' work. This would be a fund available for any snags that may arise in, say, the six months after the job is complete. If there are any problems with the building it will be paid for from that fund. If there are none, the builder can keep it. That should provide an incentive for the builder to get everything right first time, and ensure they can have no objection to resolving any issues that arise.
- Ensure any builder, plumber, or electrician, etc., is part of a professional organization.

Prepare for the complaint

If, despite your efforts, things have gone wrong and you need to complain remember Golden Rule 1: be prepared.

- Be very clear in your mind what it is you're complaining about. Complaining that a hotel is 'awful' will not get you anywhere. You need to be precise in your complaints.
- Make sure you know all the things that are wrong with a product. Have all the relevant information to hand. You'll need to know when and where you bought the product and its details.
- Think in advance about what compensation you want. Is a full refund going to cover all your costs or did you suffer other expenses as a result of the problem?

Complain politely

It's natural to feel annoyed if you think you've been sold shoddy goods or received an inadequate service. But your argument will be far more effective if you are polite. Remember that the person you're talking to is often not themselves at fault, they are merely representatives of the company. Remember the example of Jonathan earlier? He vented his anger but it got him nowhere near getting a refund.

- Using the person's name can be polite and ensure personal service. So if you're talking on the phone to a company or talking to a representative find out what their name is and use it. It helps create rapport. Show that you are a real person with real problems, not just a 'complaining customer'.
- When you come around to setting out your complaint, refer to the company. If you're talking to a representative of a bank, saying 'I feel X Bank should reimburse this charge' is far more likely to be effective than saying, 'I feel you should reimburse this charge'. You want to keep on good terms with the person you're talking to, even if you're in dispute with their company.

- Try to be positive. 'Your company has provided excellent service in the past. I'm really pleased with your product, but there's no getting away from the fact that it was late. Don't you agree that the lateness of delivery was well below the standard that you normally achieve?'

Be reasonable

Come across as being reasonable. An outrageous demand is unlikely to succeed. Asking for £3,000 to compensate you for the tummy upset caused by the prawns you bought from the local supermarket is only going to make you look silly.

Useful examples

'Of course I don't expect a full refund, because the goods are of good quality. However, I have lost money as a result of the late delivery.'

'When my groceries were delivered today there were unripe avocados in the bag. I bought them for a dinner party I'm giving tonight and now I'm stuck. I would like a refund of the cost of the avocados, please.'

Seek a result that is going to be realistic and reasonable to both sides. If you're arguing with a plumber you cannot expect him to spend 24 hours a day on your job. If you show that you recognize he has other customers you're more likely to win him around.

Make compensating you beneficial to the company

You'll make the job of complaining about the product or service easier if you can show the business they will gain from compensating you.

Useful examples

'Look, I'm always telling my friends what a great company you are to deal with. I must have passed lots of business your way, but I won't be able to talk of your company in glowing terms in the future if you're unable to recognize this wrong.'

'How about you give me £10 off my next order? Otherwise I might look to use another retailer next time.'

Do be a bit careful with the last one. If you are entitled to £10 back, accepting £10 off your next order may not be a good deal at all. Only accept a 'money back off next purchase' offer if you will definitely use it, and only if it is more than you think you will get in terms of a straightforward refund.

Making proposals that seem to evolve from the conversation rather than as a demand from you may make the request for compensation more attractive.

Useful example

'Well, I wonder if we can agree the following. The goods were unfortunately unsatisfactory and therefore Smith & Co should pay a refund. Does £60 sound like a fair figure?'

A good question to ask is: 'What is the reason why you don't want to compensate/reimburse me?' This approach operates on what lawyers call the burden of proof (we talked about this in Golden Rule 3, remember). Rather than you having to explain why you deserve compensation, the company is put in the position of explaining why they should not compensate you.

It might also be helpful to ask the company questions:

'I just want to make sure I have understood the position: do you agree that the goods were delivered late? Do you agree that as a result I lost £60 of business and suffered considerable inconvenience?'

The hope is that in answering these questions the justice of your demand becomes obvious.

Remember, you must try to understand the position of the person you're talking to. Key questions will be whether you are legally entitled to what you are claiming or whether you are asking for something as a gesture of goodwill. If the latter, you need to show the company why reimbursing you would make economic sense:

'I have had a credit card with your company for three years. I believe the charges you have put on my card are unreasonable. If you do not refund me the charges, I will change to a different credit card company.'

It can also be helpful to use 'what' questions. So rather than 'why' has a company decided something ask: '*What* are the reasons for your decision?' And rather than '*Why* cannot you deliver the goods?' ask 'Under *what* circumstances can you deliver the goods?'

Who to complain to?

Sometimes it's difficult to know who to complain to: the shop? the manufacturer? some professional organisation? There are two key factors to consider:

- Which is the most convenient for you?
- What is most likely to produce a good result?

Sometimes shops try to palm off complaints from customers by telling them to contact the manufacturer. You don't have to do this. After all, you paid your money to the shop and in law you have a contract with the shop, not the manufacturer. They can get in touch with the manufacturer if they want to. However, if the goods have failed some time after the purchase, or if what you want is a repair rather than a refund, you are likely to have more luck with the manufacturer.

A key issue is often whether the goods were defective when you bought them or whether they became defective because you maltreated them in some way. The quicker you can make your complaint the harder it will be for the shop to suggest that the problem is your fault or that the problem is 'wear and tear'. So, if something has gone wrong with a product or a service, get in touch with the provider as quickly as possible. If it's something like an unsatisfactory visit to a hotel it's best to complain there and then.

Basic legal rights

This book is not able to set out all of the relevant law on unsatisfactory sales; such a book would be enormous. And very expensive. But here are some of the main principles:

- All goods bought must comply with three requirements: they must be as described, fit for purpose, and of satisfactory quality.
- Satisfactory quality means that they reach the standard that a person would regard as satisfactory bearing in mind the price and description. The reference to price here is important. If you buy a cheap product you cannot expect the same standard as if you bought a very expensive one.
- It is the seller who is responsible if the goods do not comply with the three requirements. Purchasers are entitled to request their money back 'within a reasonable length of time'. Notice you are entitled to your money back: you don't have to accept a replacement, unless you want to.
- Generally it is the purchaser who must assume the goods were faulty.

- In the case of services, these must be provided with reasonable care and skill. If the work is not done with reasonable care and skill the work must be put right at no extra cost. If that is not done you can ask someone else to do the work and claim the cost from the original provider.
- If you paid for the goods or services by credit card you may be entitled to a refund from the credit card company if there are difficulties in dealing with the supplier of the goods or services.

Keep records

If you're embarking on a complaint it's worth keeping as much evidence as you can. Keep a copy of your letters and any replies. Write a note of any conversation you have had. Take photographs of the faulty goods or shoddy work on your property.

Go to the top

If your complaint is a serious one and you're not getting a rapid response it's worth going to the top. Write to the head of customer services and send a copy to the chief executive. The details of who these people are should be easily found on a company's website. If you don't receive an appropriate response within 14 days, write to the chief executive again.

Getting help

If you feel you need assistance, the local Citizens' Advice Bureau or the Trading Standards Officer may be able to help. If there are significant sums of money concerned using a solicitor may be necessary. Many newspapers have pages that take up readers' concerns and this might be an avenue to explore. There are television shows devoted to shoddy workmanship or faulty goods. You could even contact your MP if it is an issue that affects others as well, or if a great harm has been done to you as a result of the bad goods or service.

Another route is to complain to a professional organization. This is appropriate if you have been badly treated by someone who belongs to one of these. For example, lawyers are accountable to the Law Society.

Getting it right

Jonathan: 'Hello, is that Clogs?'

Tina: 'Good morning, sir. I'm Tina. Can I help you?'

Jonathan: 'Well, Tina, I bought a pair of shoes at Clogs yesterday and wore them for the first time yesterday afternoon. And you won't believe it, the sole came off on the first outing. I was only taking my daughter for a walk in the park.'

Tina: 'I'm sorry to hear that. Do you have the receipt?'

Jonathan: 'I do, and photographs.'

Tina: 'Well, if you can bring them in we can do a repair.'

Jonathan: 'Actually, Tina, I would rather have my money back. I'm not sure a repaired shoe is as good as a new one. I should add I'm always singing your praises to my friends.'

Tina: 'Sir, it sounds as if you are entitled to a refund. In fact, if you ask for me when you come into the shop I will deal with this myself and see if we can't give you a voucher as well as a refund.'

Jonathan: 'That sounds great.'

Summary

When complaining about goods and services be polite and firm. Get a clear picture in your mind about what is wrong and how that can be rectified. Be reasonable in your demands. When dealing with the company try to keep a good relationship with the person you're dealing with. If necessary go to the top of the firm. Don't forget there are other organizations who may be able to help you if you are not satisfied with the response from the people you have dealt with.

In practice

If you have a complaint with a business, talk to anyone you know who has been in the same position and see if you can find out helpful tips. Make sure you have an accurate record of all the details of your complaint. Keep calm and keep it all in perspective!

Chapter
15

How to get what you want from an expert

Some of the most difficult arguments arise when dealing with an expert: a head teacher, banker or doctor, for example. Inevitably you feel that they have expertise you don't and this puts you at a disadvantage. However, if they're making a decision you feel is wrong it's important to stand up for yourself. Arguing with a professional requires some specialist techniques.

Getting it wrong

Doctor: 'I'm not sure why you're here again. I told you last week that you're fine and there's nothing wrong with you.'

Sam: 'Yes, but Doctor I still feel unwell.'

Doctor: 'Well, I examined you thoroughly last week and there was nothing wrong.'

Sam: 'But I don't feel well.'

Doctor: 'I'm afraid there's nothing I can do.'

Sam: 'I feel worse than last week.'

Doctor: 'Sam, I have a lot of patients waiting for me.'

Sam: 'Well, I'd better go then.'

There are ways to get an argument with an expert right. Here are some key principles:

Respect the expert

Most experts deserve their status. But not all. Still, there's nothing to be gained by not showing due respect. Many professionals have power and considerable experience. They will be used to dealing with 'difficult customers'. Annoying them, wasting their time, or not acknowledging their expertise will not get you anywhere. No one likes being contradicted, but experts in particular are likely to be offended.

Unless they have invited you to do otherwise, call them 'Ms X', 'Mr Y' or 'Dr Z'. Never suggest you know more than they do. Make sure you arrive on time for all appointments. All of these little things will mean they're more likely to listen to you.

Prepare

Remember Golden Rule 1. Some people find seeing a doctor, lawyer or other professional intimidating and daunting. It's especially important, therefore, to think in advance about what you want to say. It may even help to write it out beforehand. It's quite common for people to visit a doctor and leave finding that somehow they have never got round to talking about the thing they were really worried about. It may even be helpful (if you're feeling very intimidated) to hand over to the professional a short note of the main points you want to raise. Indeed, they may find that the most effective use of their time.

In the case of a doctor's visit, if you have heard of other treatments for your condition, be ready to cite the source (e.g. website) of your information. When dealing with any professional, being able to substantiate what you're saying by referring to either a newspaper article or expert source gives weight to your argument.

Similarly, if you are approaching a bank for a loan, be prepared. Make sure you know your key financial details. Show that you have thought through the issue carefully as a prudent customer of the bank should!

Be concise and precise

Often the professional will want to know only the salient points of your story, not the full story. They're going to be interested primarily in the key facts, so think beforehand about what they need to know. If you're telling a doctor about a fall they don't need to know the long story of how you got to the place where you fell! Tell them what you think are the main facts. They are likely to ask you questions about the things they really need to know about.

One minute telling them the key facts and nine minutes answering their questions is likely to be a better use of time for each of you rather than ten minutes of you talking, probably with lots of irrelevant information. Try to present your information logically. If you're instructing an architect, by spending ten minutes describing your ideas and fifty minutes answering their questions you will have a more productive time than fifty minutes of you talking and ten minutes of questions.

Remember you are both experts

Just because someone is an expert in a topic does *not* mean they are an expert in everything. You are the expert in what is happening in your life. It is amazing how some people seem to think that because they're knowledgeable in one topic they can pontificate on anything. So, just because a doctor or a lawyer knows a lot about medicine or law does not mean they know everything about *you*. Indeed, remember:

You are the expert on you!

When you're talking to a doctor, the doctor knows about medicine but cannot explain how you're feeling. Your doctor might be an expert on psoriasis, but you are the expert on how psoriasis affects you.

Fortunately many doctors realize this. The old days, when the doctor told you what was wrong with you and what to do, are largely gone. Usually these days, doctors will give you information on available treatments and discuss with you what is best. Sometimes people find that disconcerting, but it normally works for the best. But hopefully you will not find a doctor like this example (a true conversation):

> **Getting it wrong**
>
> Doctor (reading case notes): 'Ah, I see you've a boy and a girl.'
>
> Patient: 'No, two girls.'
>
> Doctor: 'Really, are you sure? Thought it said . . . (checks in case notes) oh no, you're quite right, two girls.'

Likewise, when you are dealing with your child's head teacher, remember you are the expert on your child. They are an expert on education, but you know your child inside and out and can be their advocate in that situation.

Most experts deal with norms

Most experts will have standard ways of dealing with cases of a particular kind. There are general principles of guidance that they will follow in certain kinds of cases. These generally are well-proven treatments or forms of action. Normally they work well. But if you feel that the advice given is not appropriate for you, then you're going to have to explain why you are not a 'norm'. Acknowledge that for most people the advice given will be excellent, but explain why you think your case is different.

Remember, you are the expert on you. The cardiac consultant or barrister will have met you and will be dealing with you on the facts presented to them as a '31-year-old female' (or whatever). They do not know you personally. They cannot know you might not be a typical case unless you tell them. You need to explain what makes you you!

Don't be afraid to ask questions

If you're unhappy with the answers you're getting don't be afraid to ask questions, politely!

Useful examples

'Are there any other alternatives you could suggest?'

'I must admit none of the options you propose is attractive, is there nothing else?'

'Could you explain a bit more why you think that particular option is better than ...?'

Medically speaking, don't forget you have a right to refuse treatment. It's your body and you always have the right to say 'no'. If you don't feel happy about the advice given you can always tell the doctor you need time to think about it. A good doctor will respect that.

Many cases where a doctor makes a wrong medical diagnosis or a lawyer gives faulty advice flow from the patient or client not disclosing all the relevant facts. If you have concerns or issues you think are important, speak up! And try not to be embarrassed. Most doctors and lawyers have heard all sorts of bizarre things. It's better to be embarrassed and get the best advice, than unembarrassed but with bad advice.

It's very important with a doctor or lawyer that you are clear on what advice or information you have been given. There have been some terrible stories of patients incorrectly understanding how they were meant to take their medication. If you're not sure what you have been told, ask that the professional explain it again. Or maybe better, ask them to write to you with the information.

If after you get home you remember things you forgot to say that you think might be important, get back in touch with the professional. Most doctors and lawyers can be contacted by phone and so there's normally no need to make a full appointment. At worst you will be wasting their time, but by contacting them you might be avoiding a terrible decision being made.

I've focused primarily on doctors and lawyers in explaining this point, but it goes without saying that the principle applies to all professionals you deal with. Ask questions. Don't worry about looking stupid. Even if you think it might be an obvious question, if it's one that is niggling at you, ask it! In many of these relationships, you are the paying customer and therefore have the right to take as much time as you wish in getting things sorted to your satisfaction.

Check the expert

Even after you have asked questions, there's nothing wrong in asking for further information if you still feel unhappy. Any good professional will realize that bad news can be hard to accept and that hearing the same message from several sources may help.

Useful examples

'Thank you so much for explaining all of this. There's a lot to think about. Is there anything I can read about this to explain it further? Is there a good website?'

'What you've said is very disappointing news. I think it would help me to discuss this with someone else. Is there anyone you can recommend?'

Do feel free to check what the expert has said with what you can find on the internet, or by asking friends or other professionals. If your mortgage adviser has given surprising advice, get another opinion. Check mortgage and bank websites for facts.

On the other hand, don't assume that your doctor must be wrong if an internet article or what your friend's doctor recommended is different from what your doctor said to you. There may be a very good reason why your friend's situation is different from your own. But don't be afraid to go back to your doctor and respectfully ask why there is a difference if it's worrying you.

Difficult experts

So far I have assumed that your doctor, lawyer or other professional is a reasonable person. But that's not always so. There are some professionals who come across as arrogant and conceited. We're meant to feel very lucky to be honoured with their presence. Especially if you're feeling under the weather already, dealing with such a person can be difficult.

Remember that most experts at the top of their profession do not act in this way. Indeed, the fact that the person is trying to appear superior may well indicate a lack of self-confidence. It may be best in such a case to find an alternative professional. If that's not possible, here's some advice.

First, don't make the assumption that because the person has no social skills they're not good at their job. Secondly, don't take any rudeness personally. It's very unlikely your doctor or lawyer has taken a personal dislike to you. They probably treat everyone like that. That's no excuse, but it may help to accept that the person you're dealing with is a difficult person. Thirdly, don't become aggressive or arrogant yourself. Ask in a simple, calm way for the situation to be explained further. Finally, don't forget you can always ask to see another professional. There's no reason why you should put up with rudeness if it means you're not getting the service you need.

If you are stuck with a professional, you need to use your best tactics in how you go about the argument. Remember Golden Rule 3: it's not only what you say but how you say it. Reasoning with a matron at a care home that looks after your elderly relative needs careful handling. Stay calm, be reasonable, use flattery if necessary, and otherwise 'handle' the professional to get what you want.

Complaining further

Nearly all professionals belong to a professional organization. If you really are unhappy you can complain to them. At least that way you should receive some kind of explanation. However, most professional bodies will only take action against professionals where they have clearly fallen well below the expected standard.

Getting it right

Doctor: 'I'm not sure why you're here again. I told you last week that you're fine and there's nothing wrong with you.'

Sam: 'Thank you very much for seeing me again. Since last week I have had sharp pains here and here.'

Doctor: 'Well, I examined you thoroughly last week and there was nothing wrong.'

Sam: 'I understand that, but these pains are new and they're really worrying me. It would be very reassuring if you could examine me again.'

Doctor: 'All right, let's have a look.'

Summary

With professionals, remain polite and respectful. Remember that although they have expertise only you really know all there is to know about you. If you feel that the advice being given is inappropriate you need to explain why you're not a typical case. Ask questions. Check what the professional has advised against other sources. Remember, at the end of the day you must make the decision whether to follow the advice. The choice is yours.

In practice

Become adroit at using the internet to check up on information about the problems you're facing. Learn to ask questions of your professional. Make sure you understand their advice and that you have answered their questions carefully.

Chapter
16

Arguing when you know you're in the wrong

Oh dear! Our argument that seemed so convincing earlier in the day is collapsing around us. The facts we were so certain about now appear mistaken. The logic of the argument that seemed so clear is now tarnished. It's obvious that the argument is lost. It has happened to us all. What to do?

Getting it wrong

Mary: 'So, I'm afraid that Alfred's argument is based on figures that seem dubious and he has overlooked the alternative possibility that I have advocated.'

Alfred: 'Mary has completely misrepresented my argument and my points. We shouldn't listen to her.'

Mary: 'Alfred, I can go over the figures again if you like.'

Alfred: 'I think we're bored enough already.'

Mary: 'Shall we adopt my proposal? I can back up my proposal and yours has holes that would sink a ship.'

Alfred: 'No way. Mine's fine the way it is.'

Mary: 'If we adopted your proposal, the company would suffer terrible losses. You've got your figures wrong. Maybe you're not good enough to do your job properly.'

If you realize that what you have been arguing is wrong, you must be honest about it. Continuing to argue, as Alfred did, when it's clear that you're in the wrong is embarrassing. It creates a bad impression with other people. It means that any negative consequences flowing from the bad argument are likely to be exacerbated, and you will likely be held to blame.

Here are some top tips when you realize that things have gone wrong:

Stop the argument

This is essential. If you realize you have lost the argument, simply carrying on is only going to make you look stupid. You will lose respect with others and there's little to be gained. However, be sure whether you have actually lost. What might have happened is that you have lost a particular issue in the argument. This doesn't mean that you have lost the whole argument. As in a game of tennis, conceding some points gracefully does not lose the match. You'll just have to fight harder for the other points.

Accept you have lost

Having realized that you have lost the argument, the key issue is whether you should admit that you have lost, or simply change the subject. This depends on a number of issues:

- Does the issue need to be resolved? If a decision needs to be made there may be no alternative but to agree to the other person's proposal. You can do this in a way that saves face. Don't admit that you're wrong, just allow that the other person's proposal has merit.
- Is it an issue the other person feels strongly about, or was it a friendly discussion? If it's only a friendly discussion then you can throw in the towel good-naturedly. If the other person feels strongly, it's best to focus on their proposal and not discuss yours any further.
- Will you gain respect by admitting you're wrong? As odd as this might sound, sometimes people respect a person who candidly admits they got it wrong more than a person who tries to fudge a mistake. Honesty can never be underrated.

Useful phrase

'OK. You've demolished my first argument. I won't rely on that any more. But remember, I had three arguments in favour of my view and I think the other two still stand.'

Ways to end the argument

Remember Golden Rule 10. If you have lost, lose well. Of course, you can admit defeat gracefully and get on with your day.

Useful examples

'What you've said has really helped. I think you're right.'

'I now understand the situation differently. Let's do it your way.'

'I think I got it wrong. It makes sense the way you explain it.'

'Your proposal is brilliant. Let's go with it.'

But you may decide that you need to end the argument without admitting defeat. The simplest way to do so is to change the subject.

Useful examples

'Well, this is a fascinating issue but I am afraid I have to get going.'

'We could discuss this until the grass grows under our feet. But I wanted to ask you about ...'

'We must talk about this another time, but now I must get off to ...'

These are likely to be effective in ending the conversation. If the other person is insistent on you giving them some concession before you go, there are plenty of non-committal remarks you can use:

Useful examples

'Well, you've given me a lot to think about.'

'I'll need to go away and think about all of this.'

Apologizing

Sometimes after an argument you will need to apologize. Not always: losing an argument gracefully and with dignity can be a finished affair. But maybe you have done something that merits an apology. You might realize later that you behaved badly. You might have said things during the argument that you now regret.

Apologizing is extremely important.

Can you remember when a person apologized to you? How apologies apply in particular cases will vary, but thinking through how you felt when someone apologized to you can be a helpful exercise in learning to apologize yourself.

Certainly it would be counter-productive to make an excessive apology for a minor wrong. In minor cases simply saying 'I'm sorry about that' will be enough. In giving the following examples I'm assuming that something more serious has happened and a fuller apology is appropriate. Here are some key points:

- If you have time, think carefully about how to phrase the apology.
- Think about where you are properly to blame.
- An apology should recognize the hurt caused to the other person. You need to convey the fact that you have heard and understood the pain caused. If you're not sure how much that is, then ask the person.
- Accept responsibility. A proper apology will acknowledge that you are responsible for that pain. That's why the feeble apology that some politicians offer is properly acknowledged as a 'non-apology':

> 'Please accept my apologies if you were offended by what I said.'

> 'I am saddened to learn that some were upset by my comments.'

These are not proper apologies because there's no acceptance of the responsibility for causing the pain. Indeed, they could be read as suggesting that the other person is at fault for being offended!

- Where appropriate, offer an explanation. Perhaps you were particularly stressed or tired. Make it clear that doesn't mean you're not responsible for hurting them, but that you would not normally act in that way. It may be that you said something that was misinterpreted because you had not expressed yourself clearly. Explain what you meant to say and apologize for putting it in such unclear terms.
- Try to empathize with the person you are apologizing to.

Useful examples

'I know I would have been really upset if someone had said to me what I said to you. I'm really sorry.'

'I'm sure you must think I'm a really horrible person. I didn't express myself clearly because I was tired. I'm really sorry for saying that ...'

You may feel the other party has overreacted to the situation, but it's still worth acknowledging the hurt they feel.

- If appropriate, think of a way to show your apology in a practical way. Perhaps buying someone a present, arranging for a repair, taking them out to lunch, or just being kind towards them can make up for what you have done.

Remember in applying these tips what your goal is in apologizing. It is to communicate your acceptance of responsibility for the hurt you have caused and to acknowlede that you should not have behaved in that way. Ideally you want this to lead to the other person forgiving you and not holding a grudge against you. It's not simply saying sorry that matters, it's what happens as a result. Remember that you will naturally be reluctant to apologize. People hate doing it. Pride stands in the way. But it's an enormously effective tool in righting a relationship, whether it's business or personal. How many relationships have festered for want of a few words of apology?

Getting it right

Mary: 'So, I am afraid that Alfred's argument is based on figures that seem dubious and he has overlooked the alternative possibility that I have advocated.'

Alfred: 'Mary, you're right. I'm very grateful for you pointing out the doubts over my figures. I was given them by the accounts department and assumed they were correct. However, I've learned today how important it is to double check the figures.'

Mary: 'Well, thank you Alfred.'

Alfred: 'So, I can support Mary's proposal – although I wonder if there is one idea from my proposal that might be salvaged and added?'

Mary: 'That sounds interesting; do explain.'

Summary

If you realize you are wrong, think carefully whether your whole argument is mistaken or whether only a part of it is. If your whole argument has failed then stop the argument. If necessary, apologize and move on. Remember Golden Rule 10: relationships are crucial. End the argument in a way that enables the relationship to move forward in a positive way.

In practice

Remember those occasions when someone was clearly wrong and yet kept arguing. What did you think of them? Learn to apologize well. Discover what makes an apology effective. See Golden Rule 10 for advice on this.

Chapter

Arguing again and again

Do you find that you keep arguing? Perhaps it's with a particular person and every time you meet them things disintegrate into an argument. Maybe it's with a colleague who seems to oppose everything you say all the time. Or there's a particular issue in your life and whenever it comes up you find yourself losing your cool. Perhaps you've found that with your partner there is non-stop arguing and life seems a constant shouting match. Or you might be one of those parents who find that they spend more time shouting at their children than actually talking to them. What can be done?

Getting it wrong

Michael: 'How many times do I have to ask you to do the washing up?'

Tom: 'I've been so busy.'

Michael: 'Maybe today, but also yesterday and the day before. Every time I say something about it you come up with a new excuse.'

Tom: 'Look, I'm sorry.'

Michael: 'It's all well and good you saying sorry! Do something about it.'

Tom: 'OK.'

The sad thing about this argument is that one gets the impression they will be having the same argument in a few days' time.

If you find yourself in a cycle of constant arguing then here are some tips.

Avoidance

If you know there are particular issues, people or situations that always annoy you, walk away! I realize you can't walk away from some situations, but most people have issues they get worked up about that are best just avoided. Being passionate and getting into an argument over an important issue can be a good thing, but not if you find yourself arguing about it all the time. Your personal health and relationships are not worth the stress of continual argument.

Only argue if you can change something or influence someone.

Resolution

Avoidance is all well and good, but what do you do if you have a colleague, friend or partner with whom you have argued an issue through and there's still no solution. A common cause for repeated argument is that the main issue at hand is never resolved. This might seem obvious, but so often in arguments the central divisive point gets overshadowed in all the mud-slinging and verbal sparring. There's an issue of dispute, the parties disagree violently, but the matter is never put to rest. If that happens, neither party feels that the other has understood their point of view and they continue to hold a grievance against the other. Whenever they meet this unresolved issue simmers under the surface and pollutes the relationship.

For example, if you feel that a person has lied to you and you have both argued about that, then whenever you meet there will be lurking in your mind a lack of trust towards the other. In such a case it might feel like you're having a different argument with the person each time, whereas in fact it's the unresolved argument that caused the mistrust that is underlying them all.

So in repeated arguments, resolution is key. First recognize the initial issue, then work through it together in a reasoned discussion where you listen closely and find common ground. The points given in Golden Rule 9 about resolving deadlock will come in useful.

Agree to disagree

Agreeing to disagree can be a disguise for simply avoiding resolving the issue, so be careful. But it can be a useful tool for ending a repetitive argument. It works best if you can be clear on why you disagree:

Useful examples

'The reason why we disagree on whether unemployment benefits should be raised is because you think unemployed people are disadvantaged people who need help, while I think they are lazy scroungers.'

'At the heart of our disagreement is that I think we can make our car work for another year with a few repairs, while you think the car will give out completely very soon.'

If you are able to isolate the source of the disagreement and agree on it, you're better able to move on. You can then choose to agree to disagree, or to find a solution to your disagreement. Agreeing to disagree, as long as both parties can live with it, puts an argument to rest. The argument has been aired long enough; both parties have listened to the other; both are still strong in their opinion but agree to disagree. The relationship is saved and can move forward.

Humour

Sometimes humour is the best way to resolve an argument. For example, in an ongoing argument with your partner over how to fold the bath towels, show yourselves how silly you are to be getting upset over something that's so trivial when you love each other very much. Do something funny (make the towel into a headdress) or suggest an absurd solution: 'Well, the only solution will be to ask the vicar to show us how to fold towels.'

Humour can lighten the tensest of situations. It can work in the office, on the football pitch with mates, with your children, even with your cleaner. 'I know I keep asking you to hoover the cobwebs, but I don't want a haunted house theme going on in the sitting room.'

Do be careful not to use humour in a situation where the other party might think you're not taking the argument seriously. You'll have to use judgement in employing humour as a tool. But it can be incredibly effective when used adroitly.

Dead horse flogging

We all have our pet subjects that we feel strongly about and cannot understand why anyone should disagree with us. I know one person who is a passionate supporter of using the organs of anyone who has died in order to save the lives of those who need organ transplants. He simply can't understand why everyone does not agree with him. He constantly brings the subject up. There's a time where having discussed an issue several times it may be best just to put it to bed.

Useful examples

'Look, we've gone around in circles on this one before, let's leave it.'

'I think we're flogging a dead horse here. Let's talk about something else.'

Be aware, too, of your own weaknesses. You may not be able to imagine a more interesting evening than discussing, again, the arguments over capital punishment. But your friends may not share your enthusiasm. Very few people like going over and over the same issue again. They'll think you a bore, so find something else to talk about.

Is this worth it?

We partly discussed this issue in Golden Rule 2, when we looked at considering whether this is the right time and place for an argument. If you find you keep arguing about the same issue, you may want to think about whether it's really worth it. If someone keeps saying something you find annoying or insulting, is it worth picking up on it? Now, some people take pleasure in being provocative and egging on a debate. Don't be tempted, unless you enjoy it!

Asking whether an argument cycle is worth continuing is particularly important in relation to partners and children. You'll probably find yourself annoyed by lots of things your spouse or child does, but if you start to argue or complain about all of them you're going to end up stressed out, frustrated and damaging your relationship. Pick your 'fights' carefully! Just because it annoys you doesn't mean it's worth arguing about. OK, your partner keeps leaving his pants on the floor, even though you ask him not to – is it worth getting into a row about? What will you gain? And are there habits of yours he puts up with as well? Surely. So try to have some perspective on ongoing arguments and decide whether it's really worth continuing. Sometimes the advice given to parents in dealing with children (don't expect more than they can achieve, be patient, remember you're an adult) works well with spouses too!

> “Grant me the serenity to accept the things I cannot change, courage to change the things I can, and the wisdom to know the difference.” from The Serenity Prayer, used by Alcoholics Anonymous

Remember, too, that (as anyone in a long-term relationship will testify) however powerful you might think you are, you're not going to change your partner much. At least, you shouldn't expect to. Most people get stuck in their ways. You're not going to transform your scruffy girlfriend into a Hollywood film star, or your unkempt husband into a male model. Love them for who they are, not for who you want them to be.

If your recurring argument is something that causes you emotional distress and health issues, then it's worth addressing it with professional help. If it's just a pet peeve, whether it's with a loved one or a professional, then it's worth accepting that person for who they are and questioning whether your argument is worth it. Argument is a choice, and at the end of the day you need to choose what's best for you and for maintaining the relationships close to you.

If all else fails, leave

In spite of everything said in this section, you might find that the arguments keep coming. Life in the office is one long argument. In that case, it may be best to move on. Workplaces should be fun places to be. You keep arguing with your builder – find a new one. Your childminder refuses to listen – change childminders. If you cannot resolve your argument, then look for other alternatives for that situation. But before you leave, don't assume that the problems always lie with the other party. It's easy to think if there are lots of arguments that the other person is 'argumentative' and you're entirely reasonable. But, as we've seen, arguments often mask other issues. Be careful to think through the ramifications of leaving a situation, whether that means losing a friendship or suffering financial consequences.

Look before you leap.

Perhaps I could insert a cautionary note for personal relationships. Of course, if there is violence or psychological abuse, leave. But otherwise, give counselling a chance. If you've entered a relationship with someone, more than likely there was a good reason why you wanted to be with that person. You have invested a lot of time and effort in developing the relationship. Your sense of identity is partly tied to it. You have made a commitment to that person, and perhaps have responsibilities towards children. Leaving is certainly an option when things get difficult, but explore every way of reconciling the relationship first. Studies show that men who assume they will be happier after divorce usually aren't, and women who divorce usually are!

Think carefully why you keep arguing

If you find you keep arguing it's easy to assume the other person is to blame:

'Mary is so annoying.'

'People are so rude these days.'

'My spouse is so inconsiderate.'

But it usually takes two to tango. Honestly think through what triggers the arguments. Are you regularly complaining? Is there something that regularly ignites an argument? Are you arguing a lot when you're tired? Or is work stress making you argumentative? If you can find a consistent trigger for what is making you argue, then you can watch out for it. It can even become a joke: 'Oh dear, it's Monday morning: time for an argument.'

Don't let arguments get out of hand

In arguments that happen again and again, it's easy to get frustrated and for the argument to escalate into a full-blooded row. A common phenomenon is for a minor remark, one that you've heard so many times that you're sick of it, to be seized on as the basis for a row. Be very aware of how quickly a row can escalate. Act rapidly to stop it. Be very alert to a change in you or your opponent's voice indicating that a row is about to blow. Walk away. Immediately apologize for losing your temper (that doesn't mean you're conceding the point of the argument, just recognizing that the manner in which you're arguing is getting out of hand).

Things to remember

- Don't bring in arguments from the past. Focus on the issue at hand.
- Don't say things that are personal attacks and not related to the argument.
- Talk about your feelings.
- Apologize freely where appropriate.

Remember, too, how easy it is in an argument to up the ante.

Five things couples argue about

To close this chapter on how to avoid arguing again and again, I thought it would be interesting to look briefly at the common causes of arguments in relationships. Researchers have produced a list of things that couples argue about. You probably would have been able to guess these anyway. The top five are:

- money
- exes
- household tasks
- amount of time spent together
- annoyances.

If you do find you're constantly arguing about one of these things, it may be best to take some time to sit down and decide the general parameters of the issue. It's better to avoid arguing over and over again about the same thing by advance planning. For example, agree on what your weekly budget is going to be so that each of you knows what the other expects them to spend. Make a list of chores and decide on a fair distribution of them. Get your diaries together and book some time together. Discuss contentious issues that continually annoy you, such as how to share the bathroom in the morning. By working through issues, you can stop the cycle of arguing about the same thing over and over again.

Getting it right

Michael: 'How many times do I have to ask you to do the washing up?'

Tom: 'I'm really sorry, I've been busy.'

Michael: 'Maybe today, but also yesterday and the day before. Every time you come up with a new excuse.'

Tom: 'Well, Michael, I have to admit that's a fair point. I think we need to sit down and decide how we're going to divide up the household jobs.'

Michael: 'I think that's a good idea. Are you free tonight?'

Tom: 'OK.'

Needless to say, the same approach works in your relationships with your children, your in-laws, your friendships and colleagues at work.

Summary

You don't have to keep having the same argument. Get it resolved. It may be that you need to decide to ignore certain topics, or agree to disagree. It may be that you need a heart-to-heart conversation to get the issue resolved once and for all. Whatever you do, don't get stuck in the cycle of repeating the same arguments.

In practice

Ask yourself why you keep arguing. Are there triggers to avoid? Does the problem lie with you or the other person, or both? Be brutally honest when answering that question.

Chapter
18

Doormats

Are you a doormat? Do you find that you never stand up for yourself? You find yourself agreeing to things you don't want to do all the time? You do so much to avoid an argument that everyone seems to treat you as a servant? Then this is the section for you. It's time to act.

Getting it wrong

Zhu: 'Poppy, would you mind staying a bit late tonight and finishing off the project?'

Poppy: 'I was going to go out on a date with my husband, but if it's really important I'm sure we can find another evening.'

Zhu: 'Thanks, Poppy. And I wonder if, while you're around, you wouldn't mind checking the internet to find a cheap flight to the Algarve for me next week?'

Poppy: 'Oh, I suppose so.'

Zhu: 'Thanks. And you realize that the company has put a ban on overtime payments for this month?'

Poppy: 'Oh.'

Zhu: 'Thanks, Poppy, you're a saint.'

I realize many people reading this book won't have an issue with being a doormat! But believe it or not, some people do. If you are one of the readers that feel you're a doormat, you need to do something! A major problem for you will be a lack of confidence in arguing and standing up for yourself. Reading this book is a good start.

"Men are not prisoners of fate, but only prisoners of their own minds." Franklin D. Roosevelt

Are you really a doormat?

If you feel you're a doormat, think carefully if that's a fair assessment. A lot of people in work positions feel that they are doing more than anyone else, but when the figures are looked at, they are not. Many people underestimate how much work their co-workers do. But ask as honestly as you can:

- Are you putting in more hours than others?
- Are you being acknowledged for your work?
- Do others seem to get the credit for your work?
- Do you end up doing the jobs no one else wants to do?

Don't assume you're always being taken advantage of. Try to make a fair assessment. Each of the questions I just asked also applies to domestic relationships, where one partner feels they're doing more than the other in a particular area. Often we don't realize how much the other person is doing. A good way of judging equality of work in a personal relationship is whether you both have the same amount of 'free' time.

Doormats are lovely people

If you are a doormat then you're probably a very nice person. It nearly always indicates that you're a kind, sensitive individual who likes to help other people. There's a lot that's really good about being a doormat. So don't get completely down on yourself if you are one. Where problems arise, however, is when, in your enthusiasm to be helpful to others, you're not looking after yourself or those you love. In the opening scenario, Poppy was keen to help Zhu, but was she overlooking her husband? Or, indeed, depriving herself of a fun evening?

What to do if you are a doormat

Learn to say no

Elton John says that sorry is the hardest word. Perhaps, but 'no' comes a close second. I confess I used to find it difficult to say 'no'. I remember many years ago being astonished when I was invited to a party I really didn't want to go to and a friend said to me 'You could always say no'. It had genuinely not entered my mind that that was an option. It seems a lot of people don't realize that no is an option. Go on – try saying 'no': it can be fun!

Learn how to say no

If you're asked to do something you don't want to do, be honest. Explain why it is you cannot say an enthusiastic 'yes':

> **Useful example**
>
> 'I must admit I feel completely worked off my feet at the moment. I have this deadline for next week and I'm helping Steven get his report done, which is due in at the end of this week. I just don't have time to take on anything new.'

Sometimes the best way of saying no is to offer the other person a choice. If your boss asks you to take on a new task, explain that you can but you will not then have time to do a different project. Ask which she would rather you did. This can work in family situations, too:

> **Useful example**
>
> 'Darling, I can certainly write that letter for you, but it will mean I won't have time to get the dinner ready tonight. Can you organize that? Then I can write the letter.'

Being honest means making it clear you're not being lazy or seeking to shirk responsibilities. Instead, the message you convey is that you are already fully committed.

Ask yourself whether the person asking you to do the task is acting respectfully. If they are not, there's no reason why you should feel you should say yes. They may need to learn that in relationships there must be give and take. In the work environment, they need to learn to respect co-workers. If they're not respecting you, you may not be helping them by doing what they ask.

Learn to walk away

There's no reason why anyone should ever call you names or make fun of you. That's utterly unacceptable and you should not put up with it. If you find this is happening at work you should complain to someone in management about it. If it's your boss who's the problem then walk away when he does this. Ask him politely not to talk to you in that way. If there is no response, you may need to leave. If so, it might be worth seeking legal advice on whether you are entitled to compensation.

Prioritize

Remember, the problem with being a doormat is that you're too nice. You want to help everyone. But you must be honest with yourself and realize that you cannot. You don't need to feel guilty about saying no. You're probably facing many commitments and calls on your time. You can't satisfy everyone. Try to see your no in a positive light. For example: 'I've said no so that I can make sure I have enough time with my children.' If your saying no has meant that a job hasn't been done, that is the organization's problem, not yours.

Sometimes doormats feel it's selfish to refuse to help in order to do something they enjoy. That's very worthy. But remember that if you become downtrodden, dispirited and exhausted you won't be able to help anyone. Everyone needs time to themselves, if only to recharge so that they can help other people again.

It's important here to distinguish between other people's wants and true needs. A person may *want* you to do something, but that doesn't mean they *need* it done. Being a kind person, you're likely to want to meet their needs as best you can. But don't confuse that with meeting their wants. A person may *want* a gourmet dinner, but they *need* just food. Your boss may *want* you to work 12 hours a day, but the company only *needs* 8.

Avoidance – pros and cons

The temptation might be to avoid situations or people you find threatening. This can sometimes be justified, and at other times not. Are there some areas of your life where you feel more in control, and others where you feel you are a doormat? Consider carefully why this is so.

Why not offer to do extra things when you have spare capacity? That will create a good impression and make it easier when you say no. When offering extra help you can also choose the kind of jobs you like doing.

A recent survey declared that a blazing row with your boss can be very good for your heart. Men who don't complain about unfair treatment double their risk of a heart attack. That study had problems with it, but it does indicate the dangers of keeping feelings of frustration under wraps. If you think you have been badly treated at work, it's best to do something about it.

Protecting yourself

Another danger for doormats is that they tend to be especially kind to those who are rude to them. There is a hope that by being extra kind and helpful you can win over the unpleasant person. This can be a particularly destructive phenomenon in relationships where one person does everything they can to be kind to the other person to ensure they are kept pleased. Ironically, it sometimes seems that the nicer the doormat is the more unpleasant the other person becomes. This leads to the doormat being even more desperate to please. Such a cycle is a very harmful one. Friendships, partnerships and marriages should be based on equality and fairness. If yours is not, it needs to change. Your views and desires should count as much as the other person's.

When you feel that you no longer have a choice in a relationship or, at work, when you feel you have lost the power to say no, then you must act.

Wider issues

If you have concluded that you're a doormat, it's worth thinking about why this is so. Are you desperate to please people? Do you attach too much weight to other people's opinions of you and don't think highly enough about yourself? Do you like being regarded as a 'saint' who always helps out? Remember that what you think about yourself can be reflected in how others treat you. If you see yourself as weak or useless, others may see you that way, too. Conversely, if you see yourself as strong and independent, others will respect you and not 'use' you. Choose friends that make you feel good about yourself and build you up.

Getting it right

Zhu: 'Poppy, would you mind staying a bit late tonight and finishing off the project?'

Poppy: 'I'm sorry, Zhu. I'm afraid I can't do that. However, I could finish the project first thing tomorrow morning.'

Zhu: 'Oh Poppy, we really appreciate you and it would be so helpful if you could do it tonight.'

Poppy: 'Thanks, but I hope you don't mind me noting that the company doesn't seem to appreciate me enough to pay overtime!'

Zhu: 'Ah, that's true. But it would be so kind if you could just manage to stay and finish the project. Are you sure you can't?'

Poppy: 'Zhu, no I am afraid I can't. As I said, I would be very happy to put it top of my pile for tomorrow morning. I've made plans that I can't change.'

Zhu: 'OK, we'll have to make that work. See you tomorrow.'

Summary

If you're a doormat don't get too down on yourself: it reflects the fact that you are a kind person. But it might do you and your friends no good in the long run. You need to prioritize who you can help and make sure you are not improperly taken advantage of. You need to start saying no. There is plenty of advice in this chapter on how to do this.

In practice

Be honest with people when you feel overwhelmed. Don't feel that you must say yes. Make sure you have time for yourself. Then you will be better able to help other people.

Chapter

How to be a good winner

'OK, you're absolutely right. I now see how wrong I was.' For some arguers, to hear their opponent say this is the holy grail. They might feel the aim of an argument should be total defeat of the opponent and a grovelling acceptance of their brilliance. But that is rarely realistic or desirable. We talked about how to lose gracefully but positively in Golden Rule 10. It's also important to know how to be a good winner.

Getting it wrong

Viv: 'So, if you look at all that evidence you can see I'm right.'

Tom: 'Well, I can see what you mean.'

Viv: 'Come on Tom, you must now accept I'm right.'

Tom: 'Well, I suppose so.'

Viv: 'I want to hear you say, "You are right, Viv".'

Tom: 'Viv, you're always right.'

Viv: 'Seriously Tom, you must now see I'm right.'

Tom: 'Oh OK, "You are right, Viv".'

Remember Golden Rule 10: the relationship normally matters more than the argument. You can win the argument but lose the war. If the person you're arguing with is left feeling humiliated or embarrassed it's unlikely they will be keen to see you again or do business with you again. Requiring abject apologies is rarely, if ever, appropriate.

Give a way out

If it becomes clear you're winning the argument, it may be best to give the person you're talking to a way out. Don't force them into conceding your point:

Useful examples

'I really enjoyed that discussion. Shall I send you a link to that article I was telling you about so you can read it for yourself?'

'It's such a difficult issue and I often lie awake thinking about it. But perhaps we can agree that ...'

Seek agreement

Try to close an argument by referring to the agreement you have reached. The reality may be that they have come round to your point of view, or have come to accept your terms, but talking of agreement binds you both together. They are likely to feel warmer towards the argument if it's seen as the result of a mutual process.

Useful examples

'I'm glad we've found we can agree on this.'

'This is a really useful decision we've made today. Thank you for your time in discussing it.'

Involve the loser

If there has been a disagreement within a family or in a business context, try to involve the 'losers' in a positive way.

Useful examples

'Well, Lucy, we've decided to go to Alton Towers, which I know was not your first choice, but shall we say that you can choose where we'll stop for supper on the way home?'

'Tom, I realize that this plan was not your ideal, but could you take on the job of overseeing the marketing side of it?'

Lording it over the loser will get you nowhere

The temptation, having won an argument, is to lord it over the other person. But that will get you nowhere. Boasting about how clever you are and how foolish they are might feel good at the time, but you will soon be very lonely!

‘I just knew I was right and you were wrong!’

‘I'm so glad we get to do it my way, it's much better.’

Winning a one-sided victory may not be best

It's possible in some situations to argue too well. Especially in the business context, it may not be wise to end up with an agreement that is entirely in your favour. If the deal leaves the other side in a bad position they are unlikely to want to do business with you again. Or, in a personal context, if you discuss with your partner how to divide up the housework and at the end it's agreed that they'll do everything and you nothing, you'll live to regret that. The deal at the end of an argument needs to be reasonable for both sides. It needs to offer both parties something of benefit.

In conclusion, winning an argument needs to be done gracefully. I've given you the tools to win, now it's up to you to go about it in a decent and noble manner. But remember to keep all arguments in context. Pick and choose those that are worthwhile, walk away from others. Balance winning an argument against losing a relationship. Enjoy healthy arguments, but avoid destructive ones. Keep a sense of humour. Use arguments positively. They are a great tool when used properly.

Getting it right

Viv: 'Tom, this has been a very helpful discussion. Do you feel you can now support my proposal?'

Tom: 'Well, I can see it has lots of merits.'

Viv: 'I agree that it's not a straightforward issue. The concerns you have made are all valid. I just think the potential gains outweigh the risks.'

Tom: 'I think I can support your proposal now.'

Viv: 'That's excellent. Actually, I was wondering whether you would be willing to be on the committee overseeing it?'

Summary

Win well. Be generous in victory and try to move on with the person you're arguing with. Emphasize the positives for the other person from the discussion. If you win an argument make sure the other person does not leave the discussion bitter, unhappy or humiliated.

In practice

If you win an argument be quick to encourage and build up the other person. Avoid any boasting about your victory. Invite the other person to join you on a project.

Chapter 20

To recap

So now you're well equipped to succeed in arguing. Let's finish with a recap of our *Ten Golden Rules*.

1. **Be prepared**. Make sure you know the essential points you want to make. Research the facts you need to convince your opponent.
2. **When to argue, when to walk away**. Think carefully before you start to argue: is this the time; is this the place?
3. **What you say and how you say it**. Spend time thinking about how to present your argument. Body language, choice of words and manner of speaking all affect how your argument will come across.
4. **Listen and listen again**. Listen carefully to what the other person is saying. Watch their body language, listen for the meaning behind their words.
5. **Excel at responding to arguments**. Think carefully about what arguments the other person will listen to. What are their preconceptions? Which kinds of arguments do they find convincing?
6. **Watch out for crafty tricks**. Arguments are not always as good as they first appear. Be wary of your opponent's use of statistics. Keep alert for distraction techniques such as personal attacks and red herrings. Look out for concealed questions and false choices.
7. **Develop the skills for arguing in public.** Keep it simple and clear. Be brief and don't rush.
8. **Be able to argue in writing**. Always choose clarity over pomposity. Be short, sharp and to the point, using language that is easily understood.
9. **Be great at resolving deadlock**. Be creative in finding ways out of an argument that's going nowhere. Is it time to look at the issue from another angle? Are there ways of putting pressure on so that the other person has to agree with you? Is a compromise possible?

10. **Maintain relationships.** This is absolutely key. What do you want from this argument? Humiliating, embarrassing or aggravating your opponent might make you feel good at the time, but you might have many lonely days to rue your mistake. Find a result that works for both of you. You need to move forward. Then you will be able to argue again another time!

Index

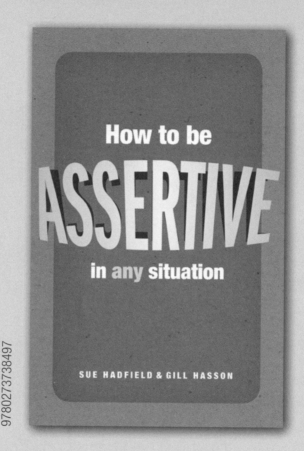

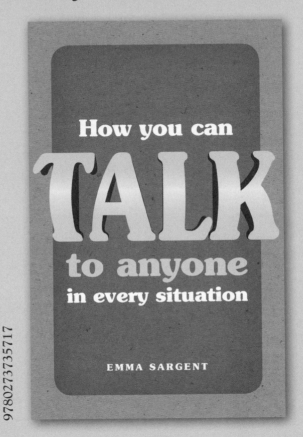